T0092736

5G and Beyond

5G and Beyond

The Future of IoT

Edited by
Parag Chatterjee
Robin Singh Bhadoria
Yadunath Pathak

CRC Press
Taylor & Francis Group
Boca Raton London New York

CRC Press is an imprint of the
Taylor & Francis Group, an **informa** business

A CHAPMAN & HALL BOOK

First edition published 2022
by CRC Press
6000 Broken Sound Parkway NW, Suite 300, Boca Raton, FL 33487-2742

and by CRC Press
2 Park Square, Milton Park, Abingdon, Oxon, OX14 4RN

CRC Press is an imprint of Taylor & Francis Group, LLC

ISBN: 978-0-367-49329-5 (hbk)
ISBN: 978-0-367-49336-3 (pbk)
ISBN: 978-1-003-04580-9 (ebk)

DOI: 10.1201/9781003045809

Typeset in Times LT Std
by SPi Technologies India Pvt Ltd (Straive)

Access the companion website: https://www.routledge.com/9780367493295

Contents

PART I Fundamental Architectural Concepts for 5G and IoT

PART II Applied Scenarios of 5G and IoT

Preface

5G technology is acting as a fundamental catalyst for the paradigm shift of mobile communications in Internet of Things (IoT) devices across all the service sectors. The IoT is unveiling its ubiquitous presence in contexts spanning from smart transport to smart healthcare, from smart education to smart shopping. The global craving for smart, connected, and intelligent services has ushered in newer concepts in IoT, fortifying its transdisciplinary reach through domains like sensors and network technologies, internet protocols, cloud and edge computing, big data analytics, artificial intelligence, data security, and privacy.

The rollout of 5G across the different countries of the world raises newer perspectives toward its integration with IoT. For IoT-based smart devices, 5G not only means speed, but also better stability and efficiency, and more secure connectivity in multifarious areas like self-driving vehicles, smart grids for renewable energy, intelligent healthcare services, and so on.

This book presents a two-fold approach: on one hand it discusses the fundamentals and theoretical concepts of 5G with respect to the IoT, and on the other hand it highlights some applied research on allied areas of 5G and IoT. Thus, in addition to providing a landscape view of recent trends in the field of 5G pertaining to IoT, this book also opens up discussion on future trends in mobile networking for IoT–5G and beyond.

Parag Chatterjee

Robin Singh Bhadoria

Yadunath Pathak

Editors

Parag Chatterjee is a researcher at the National Technological University *(Universidad Tecnológica Nacional)* in Buenos Aires, Argentina and an assistant professor in the Department of Biological Engineering, University of the Republic *(Universidad de la República)*, Uruguay. He received his MSc in Computer Science in 2015 from University of Calcutta, and is currently working in the transdisciplinary areas of the Internet of Things and artificial intelligence applied to healthcare, especially in the domain of intelligent prediction and prevention of cardiometabolic diseases. He is a member of the editorial and review boards of more than 20 international journals and conferences, and has delivered talks and keynote speeches at several international conferences and events, including ExpoInternet LatinoAmérica, IoT Week Geneva and TED^x.

Robin Singh Bhadoria is currently working as an assistant professor in the Department of Computer Engineering & Applications at GLA University, Mathura, Uttar Pradesh, India. He has more than ten years' experience in teaching as well as research. Dr Bhadoria received his PhD in Computer Science and Engineering from the Indian Institute of Technology (IIT) Indore, Madhya Pradesh, India in 2018. He received his Master's of Technology and Bachelor's of Engineering degrees in Computer Science and Engineering from Rajiv Gandhi Proudyogiki Vishwavidyalaya, Bhopal, India, securing the University Gold Medal in 2011. He is a professional member of IEEE (USA), the Internet Society, the Institution of Engineers Kolkata, and has edited more than seven books published by international publishers including CRC Press (Taylor & Francis), Springer, and IGI Global. He has published over 40 articles in reputed international journals and presented more than 21 peer-reviewed papers at international conferences, including presentations in Italy, Russia, and Serbia. He is currently working in the domain of the Internet of Things.

Yadunath Pathak is an assistant professor at the Indian Institute of Information Technology (IIIT), Bhopal, India. He received his PhD and Master's of Technology degrees from ABV Indian Institute of Information Technology & Management (IIITM), Gwalior, India. He completed his Bachelor's of Engineering (BE) degree in Information Technology at RGPV, Bhopal, India in 2009. He is currently working in the domain of machine learning and data analytics and has authored over 30 articles for international journals and conferences.

Contributors

Taniya Anand
Department of Computer Science and
  Engineering, Women's Institute of
  Technology
Dehradun, Uttarakhand, India

Ashish Bagwari
Department of Electronics and Communication
  Engineering, Women's Institute of
  Technology
Dehradun, Uttarakhand, India

Jyotshana Bagwari
Department of Research and Development,
  Robotronix Engineering Tech Pvt. Ltd.
Indore, India

Robin Singh Bhadoria
Department of Computer Engineering &
  Applications, GLA University
Mathura, Uttar Pradesh, India

Ankit Bhurane
Department of Electronics and Communication
  Engineering, Visvesvaraya National Institute
  of Technology (NIT)
Nagpur, Maharashtra, India

Arun Chauhan
School of Computer Science & Engineering,
  University of Petroleum and Energy Studies
  (UPES)
Dehradun, Uttarakhand, India

Brijesh Kumar Chaurasia
Department of Information Technology, Indian
  Institute of Information Technology (IIIT)
Lucknow, Uttar Pradesh, India

Laasya Cherukuri
Department of Electronics and Communication
  Engineering, Indian Institute of Information
  Technology (IIIT)
Nagpur, Maharashtra, India

Ram Kishan Dewangan
Department of Electronics and Communication
  Engineering, Thapar Institute of Engineering
  and Technology (TIET)
Patiala, India

R.P.S. Gangwar
Director, Women's Institute of Technology
Dehradun, Uttarakhand, India

Shruti Goel
Department of Electronics and Computer
  Engineering, Institute of Technology, Nirma
  University
Ahmedabad, Gujrat, India

Mohammad Kamrul Hasan
Center for Cyber Security, Faculty of
  Information Science and
  Technology, Universiti Kebangsaan
  Malaysia (UKM)
Bangi, Malaysia

Shah Mahdi Hasan
School of Engineering, The University of
  Newcastle
Australia

Md Mashud Hyder
School of Engineering, The University of
  Newcastle
Australia

Ashwin Kothari
Department of Electronics and Communication
  Engineering, Visvesvaraya National Institute
  of Technology (NIT)
Nagpur, Maharashtra, India

Malay Kumar
Department of Computer Science and
  Engineering, Indian Institute of Information
  Technology (IIIT)
Dharwad, Karnataka, India

Rakesh Kumar
Department of Computer Science and
    Engineering, National Institute of
    Technology (NIT)
Hamirpur, India

Kaushik Mahata
School of Engineering, The University of
    Newcastle
Australia

Svetislav Maric
University of California San Diego, UCSD
    Extension
San Diego, CA, USA

Varun Mishra
Department of CSE, Amity School of
    Engineering & Technology, Amity
    University
Gwalior, Madhya Pradesh, India

Tushar S. Muratkar
Department of Electronics and Communication
    Engineering, Indian Institute of Information
    Technology (IIIT)
Nagpur, Maharashtra, India

Priyanka Pateriya
Department of Electronics and
    Communications Engineering, Rajiv Gandhi
    Proudyogiki Vishwavidyalaya (RGPV)
Bhopal, Madhya Pradesh, India

Yadunath Pathak
Department of Information Technology, Indian
    Institute of Information Technology (IIIT)
Bhopal, India

Meena Pundir
Chitkara University Institute of Engineering
    and Technology, Chitkara University
Punjab, India

Deepak Rai
Department of Computer Science and
    Engineering, National Institute of
    Technology (NIT)
Patna, India

Milind Raj
Department of CSE, Indian Institute of
    Information Technology (IIIT)
Bhopal, India

Shyam Singh Rajput
Department of Computer Science and
    Engineering, National Institute of
    Technology (NIT)
Patna, India

Jasminder Kaur Sandhu
Chitkara University Institute of Engineering
    and Technology, Chitkara University
Punjab, India

Rohan Sharma
Department of CSE, Indian Institute of
    Information Technology (IIIT)
Bhopal, Madhya Pradesh, India

Anurag Shukla
Department of Computer Science and
    Engineering, National Institute of
    Technology (NIT)
Raipur, Chhattisgarh, India

Piyush Shukla
Department of Computer Science and
    Engineering, Rajiv Gandhi Proudyogiki
    Vishwavidyalaya (RGPV)
Bhopal, Madhya Pradesh, India

Ranjana Sikarwar
Department of Computer Science and
    Engineering, ASET, Amity University
Gwalior, Madhya Pradesh, India

Anuradha Singh
Department of Basic Sciences, Indian Institute
    of Information Technology (IIIT)
Nagpur, Maharashtra, India

Mangal Singh
Department of Electronics and
    Telecommunication Engineering, Symbiosis
    Institute of Technology, Symbiosis
    International (Deemed University)
Pune, India

Jyoti Singhai
Department of Electronics and Communication
 Engineering, Maulana Azad National
 Institute of Technology
Bhopal, Madhya Pradesh, India

Rakesh Singhai
Department of Electronics and
 Communications Engineering, Rajiv Gandhi
 Proudyogiki Vishwavidyalaya (RGPV)
Bhopal, Madhya Pradesh, India

Prateek Srivastava
Chitkara University Institute of Engineering
 and Technology, Chitkara University
Punjab, India

Ayushi Tandon
Department of Electronics and Communication
 Engineering

Indian Institute of Information Technology
 (IIIT)
Nagpur, Maharashtra, India

Sarsij Tripathi
Department of Computer Science and
 Engineering,
 Motilal Nehru National Institute of
 Technology (MNNIT)
Allahabad, Prayagraj, Uttar Pradesh, India

Lazar Z. Velimirovic
Mathematical Institute SANU
Belgrade, Serbia

Parth Wazurkar
Department of Computer Science and
 Engineering, Indian Institute of Information
 Technology (IIIT)
Nagpur, Maharashtra, India

Part I

___

Fundamental Architectural Concepts for 5G and IoT

1 The Impact of Artificial Intelligence on 5G-Enabled IoT Networks

Ranjana Sikarwar
Amity University, Gwalior, India

Parth Wazurkar
Indian Institute of Information Technology (IIIT) Nagpur, Nagpur, India

CONTENTS

1.1 INTRODUCTION

The Internet of Things (IoT) technology, constantly growing to accommodate the needs of future IoT applications, currently uses existing 4G networks. IoT-centric applications such as augmented reality, which work through intelligent connections, require higher data rates, large bandwidth, increased capacity, low latency and high throughput in order to function quickly [15]. The exponential growth of IoT devices that generate massive amounts of data wirelessly has led to widespread investigation of high speed, enhanced bandwidth and low latency 5G cellular networks. 5G is widely expected to achieve enhanced end-user quality of experience (QoE) and higher data rates than 4G, as well as 1000 times greater system throughput and ten times greater spectral efficiency. Current IoT arrangements face many technical challenges, including security, compatibility and longevity, the large number of node connections and new standards. Higher data rates, low latency, efficient use of the spectrum and seamless connectivity between different networks are the most debated topics in IoT [15, 21].

Artificial intelligence (AI), with its basis in the disciplines of computer science, mathematics, and engineering, is a concept that simulates human intelligence in machines, using new methods, theories and the latest technological application systems. AI is used to analyze the bulk data produced by numerous IOT devices and make intelligent decisions accordingly. The aim of AI research is to solve real-world complex problems without human intervention so that machines can replace humans. Among the useful tasks that can be performed by smart systems using AI technology are diagnosing diseases such as cancer more effectively than doctors, or safe driverless vehicles. Artificial intelligence can be used to develop smart systems for office, business or home use [9, 19].

AI aims at developing an automated computer with software enabling it to think intelligently in the same way as humans can reason in an intelligent way. Machines can also learn and solve problems in an intelligent manner, hence the terms 'machine learning' and 'machine intelligence'. Artificial intelligence has made real progress in numerous areas such as image recognition, natural language processing, medical image analysis, game playing and many more [14].

1.1.1 Artificial Intelligence: State of the Art and Prospects

Before developing AI programs, we need to know what intelligence is. Intelligence comprises perception, reasoning, learning, problem -solving, and so on. Figure 1.1 uses a Venn diagram to show elements of intelligence, with intelligence in the middle, surrounded by learning, perception, reasoning and problem solving.

Thus, the aims of program development are:

- To correlate perception and action smartly with huge knowledge of the world
- To know how to solve a real-world problem accurately and using cognitive functions
- To apply high-level logic to solution finding,
- To solve knowledge-based tasks and learn from mistakes
- To mimic human intelligence.
  - a. *Perception:* The process of inferencing, acquiring, selecting, and organizing valuable information from unprocessed input. Humans use their experience, sense organs and the environment for perception. Intelligent machines undertake perception in a logical manner [3].
  - b. *Learning:* The process by which new understanding can be acquired from different sources, such as books, life experience, teaching by experts, knowledge and skills gained through study. Learning increases a persons knowledge in new fields and areas.

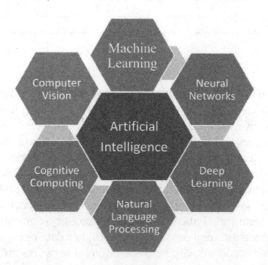

FIGURE 1.1 Elements of intelligence.

    c. *Reasoning:* The act of thinking in a logical way to predict something, to judge or make decisions on real-world cases [1].

    d. *Problem solving*: The method of discovering a problem, analyzing it, then finding the best solution in the minimum time using generic or ad hoc methods [20].

1.1.2 IMPORTANT SUBSETS OF AI

The subfields of AI (see Figure 1.2) make machines work smarter, ranging from estimating the price of your ride in car booking apps through to self-driving cars [5]. These subfields are contributing to 5G by helping wireless networks to become proactive and predictive in nature.

Machine learning (ML) uses different algorithms to make predictions from data. The 5G infrastructure will include many networks operating at different frequencies resulting in complex scenarios with multiple variables. ML tools will play a key role in outlining the patterns in these scenarios and finding unknown patterns by using machine-learning algorithms [14]. A principal use is image analysis in healthcare, including for the diagnosis of diseases. Popular machine-learning tools used today include Python, R, MATLAB, Spark and TensorFlow [16].

Deep learning learns all the human characteristics and behavioral databases to carry out supervised learning. Models, images, text or sound are classified through the processing and analysis of input data using different methods or algorithms to get the desired output. The classification or grouping of similar datasets (such as images or documents) to predict future events is known as predictive analysis. Among the popular algorithms used are neuro-evolutionary methods and gradient descent [6]. Models are trained continuously with huge labeled datasets along with multi-layered neural network architectures.

Artificial neural networks (ANNs) are the real brains of AI. They help to process huge bytes of data using edge computing in 5G networks. ANNs consist of layers of neurons called perceptron which are biologically inspired programming concepts like human brains. The ANN is trained with many training examples for the desired output. Deep learning with neural networks can be very useful in many areas like image processing, video processing, natural language processing, etc. The network is trained on a specific group of attributes. The nodes lying deeper inside the layers of the neural network, identify the more complex attributes, resulting in a final output.

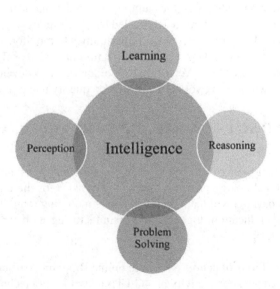

FIGURE 1.2 Subfields of AI.

Cognitive computing is the simulation of human thought processes to solve complex tasks. This includes speech and vision recognition, machine learning, reasoning, human computer interaction, etc. Interestingly, the machines tend to visualize and grasp human behavior, understand sentiments in different situations and simulate human thought processes [7].

In *natural language processing*, computers can understand, identify, locate, and process human language (text and spoken words) and speech. This subfield of AI enables computers to interact with machines and human language to deliver logical responses [7, 17].

Computer vision is a field of computer science that makes computers recognize, analyze, and interpret real-world images using a combination of deep learning and pattern recognition [1, 2].

1.1.3 BACKGROUND OF AI

Neurology research considers the brain to be an electrical network of neurons. Claude Shannon's research in information theory was illustrated by digital signals 0 or 1. Norbert Wiener's cybernetics produced the theory of control and stability in electrical networks. In the year 1943 Walter Pitts and Warren McCulloch described how artificial neural networks could perform simple logical functions. Their work on artificial neural networks led to the development of what researchers now call neural networks. In 1950 Alan Turing's landmark paper on creating machines that can think introduced the famous Turing Test. In 1956, at the Dartmouth Summer Research Project, the field of AI was founded; artificial means 'man-made' and intelligence means 'thinking ability', thus AI means 'man-made thinking ability. Since 2011, advances in deep learning have promoted research in the field of image and video processing, and text analysis.

1.1.4 CURRENT RESEARCH IN 5G

A heterogeneous IoT network incorporates 2G/3G/4G, Wifi, Bluetooth and other technologies to connect billions of devices wirelessly. For many years 2G networks have been used for voice communication, 3G services for both voice and data, and 4G for broadband internet. Since 2012, 4G and 4G LTE (long-term evolution) services have been widely used for broadband internet services. 3G and 4G are not fully optimized technologies for IOT networks due to challenges such as complex communication, device computational capabilities, smart decisions etc.

Current research in 5G networks uses MIMO (multiple-input-multiple-output) antennas to handle manifold data conversations talking on the same data signal at the same time. AI and machine learning algorithms handle all network traffic controlled by 5G by analyzing historical data patterns and conversations. AI and ML algorithms provide opportunities to monitor, optimize or manage the network and allow owners to control their resource utilization in the network.

Technology giants Intel, CISCO and Verizon have collaborated to develop 5G-based neuroscience-based algorithms for wireless networks to adapt video quality using intelligence.

1.2 ROLE OF AI AND 5G IN DIGITAL TRANSFORMATION ACROSS INDUSTRIES

The evolution of the digital world from 4G to 5G wireless technology has offered an enhanced level of connectivity. 5G technology provides faster data transfer for the massive amounts of data generated by users every day. 5G will help many industries including retail, transport, supply-chain management, the military, education, logistics and manufacturing with rapid business operations and reliable connectivity [20].

- *Retail*. Nowadays millions of people engage in online shopping, particularly on smartphones, due to the high-speed data connectivity on 4G/LTE networks. 5G technology will enhance the speed and utility of e-commerce drastically.

- *Transport*. 5G wireless networks will help connect both public and private vehicles with real-time access with an enhanced level of visibility and precision.
- *Supply-Chain Management*. The combined benefits of 5G, AI and IoT have improved the functioning of supply chain management. IOT-based data sensors ensure real-time availability among stakeholders. AI methods help supply chain management to take intelligent decisions and accurately predict production demands.
- *Military*. The military always needs continuous uninterrupted mobile device communications. 5G wireless technology will play an important role in the use of missiles traveling at a greater speed than sound. AI and IoT can collect data about material usage at military sites and inform optimum spending and resource utilization.
- *Education*. Virtual reality (VR) and augmented reality (AR) can help students in interactive classroom teaching with virtualized 3D models. 5G fulfills the significant demands of AR/VR applications in terms of bandwidth and latency.

1.3   IMPACT OF MACHINE LEARNING FOR A 5G FUTURE

The 5G setup will require the use of different frequencies and other services resulting in an intricate and multivariable scenario. Various tools are needed according to the service demand for the network setup and planning (e.g., height of antennas, power required and volume of the network) for the 5Genabled IoT ecosystem.

ML uses the large amount of data generated by 4G infrastructures to predict the behavior of device users and predict patterns.

The two methods of learning based on algorithms are:

- *teaching supervision*, which teaches the AI how to behave and; to learn a new behavior following result analysis;
- *reinforcement by merit*, in which AI is rewarded for carrying out the assigned task.

1.3.1   CATEGORIZATION OF MACHINE LEARNING MODELS FOR 5G DEPLOYMENT

Machine learning in 5G is subdivided into supervised, unsupervised, and reinforcement learning. The ML model used will vary in each phase of planning and deployment work for 5G. A detailed block diagram of ML models is shown in Figure 1.3.

Supervised Learning. Supervised models can be used to predict the values learnt from previous models, such as channel capacity or radio propagation. In supervised learning data is provided in the form of examples with labels that predict the classified or labeled output. This is analogous to teaching a child with flash cards. The input data acts as training data, trains the model or machine and uses this training to predict future events of any new input within the known classifications. In this algorithm learning is guided by a teacher which is itself a dataset [11]. When fully trained, the supervised learning algorithm will output a new, never-before-seen example and predict a good output. Image classification and market prediction are examples of this type of learning.

Unsupervised Learning. In unsupervised learning we have no idea of the final outputs. A 5G model is planned to predict network behavior patterns from the big data obtained from IoT devices. The algorithms will be given a lot of data and tools to understand the properties of that data. A function is generated to describe completely hidden and unlabeled patterns. Gradually this algorithm learns to group or cluster the data so that if new data is input it can be classified into the right cluster. Examples are clustering of similar data and easy visualization of high dimension data.

Reinforcement Machine Learning. In reinforcement learning the system learns from mistakes, learns ways to optimize selected variables and learns from outputs obtained in the past based on pre-existing conditions [14]. Bayesian networks can be developed using this kind of learning to predict the most likely conduct of the designed network. This type of learning is used in video games such

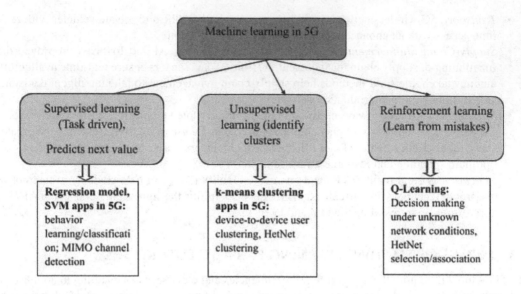

FIGURE 1.3 Machine-learning types and standards in 5G.

as Alpha Zero, Mario and; AlphaGo, as well as in industrial simulation and resource management, for example Google's data centers [21].

1.4   POTENTIAL AND LIMITATIONS OF AI AND MACHINE LEARNING FOR 5G

1.4.1   POTENTIAL OF AI

The introduction of new technologies like AI and ML to the complex 5G network and other real-world applications brings with it both advantages and disadvantages.

Artificial intelligence is difficult for beginners, yet it offers great opportunities for developing intelligent machines that can transform computer science.

- Intelligent systems reduce human error and can perform different tasks with greater accuracy and precision.
- ML may support massive machine-type communication connecting a large number of wireless devices to ensure scalable connectivity in 5G networks.
- ML provides enhanced mobile broadband (eMBB) for applications requiring higher data rates.
- AI robots or intelligent systems can easily perform many difficult tasks which carry risk and security implications for humans, for example, coal or oil mining bomb disposal and underwater exploration.

1.4.2   LIMITATIONS OF USING AI AND ML

The drawbacks of artificial intelligence include:

- The overall cost of implementing AI products is very expensive. Only a few organizations can afford it. Small organizations are not able to purchase the high-end machines, software, and resources required.

- ML algorithms only work well with high quality data. Distinguishing labeled from unlabeled data is important when deciding the type of learning to be used.
- Incorrect selection of hyperparameters (values set in ML) may affect the values obtained as a result of learning.
- Artificial intelligent systems cannot make decisions. Robots do not have the capacity to determine what is right or wrong.
- AI machines are unable to think out of the box and only perform tasks for which they are programmed. For anything else they may give false results.

1.5 REQUIREMENTS AND KEY ENABLING TECHNOLOGY IN 5G IoT

Table 1.1 shows the expectations of 5G systems and how they will be met. The principal technologies and approaches required for such systems are discussed below:

- *Wireless Network Virtualization (WNV):* Virtualization of wireless resources has become the principal trend in 5G networks. WNV ensures high resource utilization and jointly with C-RAN architecture shares resources (e.g., network infrastructure, licensed spectrum, etc.) among multiple operators [22].
- *MIMO:* Multiple-input multiple-output (massive-MIMO) technology plays an important role in attaining higher spectrum efficiency and the use of different band millimeter-wave (mm-wave) communications technologies using 28GHZ and 38GHz bands.
- *C-RAN*: Cloud-based radio access network (C-RAN) balances the load and supports MIMO concepts, mitigating the cost of baseband processing and reducing energy consumption [22].
- *D2D communication*: The volume of existing cellular networks is increased by incorporating numerous small cells (femtocells and picocells). The provision of peer-to-peer (P2P) communication (e.g., device-to-device (D2D) and machine-to-machine (M2M) communication) enables multi-tier heterogeneous networks.
- *Full duplex (FD):* Full Duplex simultaneous transmission and reception (e.g., full duplex communication) provides higher data rates. Prevailing wireless networks issues, such as long end-to-end delays, hidden terminal problem and low throughput of the network due to congestion, can thus be solved.

TABLE 1.1
5G Expectations and Proposals

Expectations and Proposals of 5G Networks

5G Expectations	Proposals
Higher data rates and throughput improvement (cell data rate 10Gb/s; throughput improvement of 1000x over 4G)	Multi-tier network, D2D communication, massive-MIMO, spectrum reuse, C-RAN
Enhanced energy efficiency (long battery life)	Energy harvesting, wireless charging
Mitigation in latency (2 to 5 ms end-to-end latencies)	Full-duplex communication, C-RAN
Increase in network density	Heterogeneous and multi-tier network
Advanced applications and services (e.g., smart city and service-oriented communication)	M2M communication, network virtualization
Internet of things, self-governing applications and network management	Self-organizing and cognitive network, M2M communication

1.6 ARTIFICIAL INTELLIGENCE DRIVEN CASES FOR REAL-TIME BUSINESS AND 5G IoT

1.6.1 COVID-19, DIGITAL HEALTHCARE AND THE ROLE OF 5G

Technology has been playing an important role facilitating digital health services during the COVID-19 pandemic. New healthcare services or applications are being combined with the latest technologies including AI and ML in the data analysis process. Machine type communication (MTC) services ensure the connectivity of different types of IoT devices like sensors, wearables and robots, enabling a COVID-19 patient to be monitored remotely. Low Power wearable devices worn by patients transfer data via Bluetooth Low Energy (BLE) which is rapidly updated to the cloud using the 5G network. The sensors in the wearable device also regularly send signals requiring action to the remote nurse through the 5G network.

1.6.2 REAL-WORLD BUSINESS USE CASES FOR AI

AI is used in many sectors, including finance, education, healthcare, banking, retail, energy, utilities, and technology [18, 20].

Healthcare: AI has become an integral part of healthcare and in many cases, AI makes a better and faster diagnosis than humans.

Social media: Social media websites like Instagram, Facebook, Twitter, and many others produce massive amounts of data that needs to be stored and managed very efficiently. AI can structure and manage data using machine learning algorithms. AI can use deep learning to detect facial features, identify the latest trends, and hashtags, and the requirements of different users. Machine learning can also be used to personalize the user's feeds according to their interests.

Education: Artificial intelligence is providing great help to teachers with automated grading, freeing up time for teaching. An AI chatbot can provide teaching assistance to students. In the near future AI will become a personal virtual tutor for students, available anytime, anywhere.

Robotics: Robotics is a multi-disciplinary domain of engineering which combines mechanical, electrical and computer engineering. Humanoid robots can talk and behave like humans. Examples of intelligent robots that behave like humans are Erica and Sophia. Using AI, we can create intelligent robots that can perform tasks without being pre-programmed, using their own experience.

Entertainment: AI based applications used largely as entertainment services include Netflix and Amazon. ML/AI algorithms help with content personalization.

Agriculture: Agriculture is becoming digital with the help of AI. Applications of AI boost the processes of sowing, growing, harvesting and sales. A number of AI services like predictive analysis; and computer vision, coupled with IOT devices, can significantly enhance productivity of crops.

Astronomy: Artificial intelligence has a remarkable role to play in contemporary space exploration. AI technology is used to understand the origin of the universe and solve complex problems. AI and machine learning are used to operate satellites and to extract maximum information from space images.

Finance: AI and the financial industries are well matched. Machine learning helps the financial sector in many ways ranging from approving loans to credit scores and managing assets. Chatbots, adaptive intelligence, and machine learning are being incorporated into financial processes.

Data Security: AI can be used to improve company data security and protect against cyber-attacks. AI software such as AEG Bot, or AI2 Platform, finds software bugs and cyber-attacks in company systems.

E-commerce: AI has lent a competitive edge to the e-commerce industry, providing many benefits to customers such as predictive product recommendations, dynamic pricing and personalization techniques that help users to explore associated products of recommended size, color, or even brand.

Travel and Transport: AI plays a vital role in the travel industry. AI can provide online customer service with chatbots or face-to-face interaction by robots for fast and better service. AI also helps make travel arrangements by suggesting hotels, flights, and routes to customers.

Automotive industry: The automotive industries are working on developing self-driving cars. Virtual assistants such as Tesla's Tesla Bot help improve performance.

1.7 CONCLUSIONS

This chapter focuses on the impact and use of artificial intelligence and its subfields such as machine learning on the 5G enabled IoT ecosystem. With an increasing number of internet of things devices and applications, there is a need for high bandwidth and larger data rates. AI and 5G will help IoT networks to take a big leap forward. This chapter also discusses technologies such as mm-wave, massive-MIMO and C-RAN which will have an impact on the design and development of 5G networks.

REFERENCES

1. "What Is Artificial Intelligence: Definition & Sub-Fields Of AI", June 3, 2020. Accessed on: June 3, 2021. [Online]. Available: https://www.softwaretestinghelp.com/what-is-artificial-intelligence
2. "The Present and Future of Computer Vision", June 26, 2019. Accessed on: June 3, 2021. [Online]. Available: https://www.forbes.com/sites/cognitiveworld/2019/06/26/the-present-and-future-of-computer-vision
3. "Artificial Intelligence Wiki", June 3, 2021. Accessed on: June 3, 2021. [Online]. Available: https://en.wikipedia.org/wiki/Artificial_intelligence
4. "Artificial Intelligence - Overview", June 3, 2021. Accessed on: June 3, 2021. [Online]. Available: https://www.tutorialspoint.com/artificial_intelligence/artificial_intelligence_overview.htm
5. "What is Deep Learning - 3 Things You Need to Know", June 3, 2021. Accessed on: June 3, 2021. [Online]. Available: https://in.mathworks.com/discovery/deep-learning.html
6. "Cognitive Computing", June 3, 2021. Accessed on: June 3, 2021. [Online]. Available: https://www.cognizant.com/glossary/cognitive-computing
7. "New AI Computer Vision System Mimics How Humans Visualize and Identify Objects", June 3, 2021. Accessed on: June 3, 2021. [Online]. Available: https://www.sciencedaily.com/releases/2018/12/181220163210.htm, 2018
8. "The Major Goals and Fields of Artificial Intelligence", March 11, 2021. Accessed on: June 3, 2021. [Online]. Available: https://www.profolus.com/topics/the-major-goals-and-fields-of-artificial-intelligence
9. "What is Machine Learning?" June 3, 2021. Accessed on: June 3, 2021. [Online]. Available: https://www.ibm.com/cloud/learn/machine-learning,202015151
10. Chunxiao Jiang, Haijun Zhang, Yong Ren, Zhu Han, Kwang-Cheng Chen, Lajos Hanzo. 2017. "Machine Learning Paradigms for Next - Generation Wireless Networks", *IEEE Wireless Communications*, 24(2), 98–105.
11. "Big Data vs. Artificial Intelligence", May 30, 2018. Accessed on: June 3, 2021. [Online]. Available: https://www.datamation.com/big-data/big-data-vs.-artificial-intelligence.html
12. "Types of Machine Learning", June 3, 2021. Accessed on: June 3, 2021. [Online]. Available: https://www.educba.com/types-of-machine-learning
13. "What are the Advantages and Disadvantages of Artificial Intelligence", June 24, 2020. Accessed on: June 3, 2021. [Online]. Available: https://www.edureka.co/blog/what-are-the-advantages-and-disadvantages, 2020
14. IEEE Staff. 2018. *2018 ITU Kaleidoscope Machine Learning for a 5G Future (ITU K)*. IEEE.
15. Kinza Shafique, Bilal A. Khawaja, Farah Sabir, Sameer Qazi and Muhammad Mustaqim. 2020. "Internet of Things (IoT) for Next-Generation Smart Systems: A Review of Current Challenges, Future Trends and Prospects for Emerging 5G IoT Scenarios", *IEEE Access*, 8, 23022–23040.
16. "The Future of Learning", June 3, 2021. Accessed on: June 3, 2021. [Online]. Available: https://www.teachthought.com/the-future-of-learning/10-roles-for-artificial-intelligence-in-education

17. Dinesh G. Harkut and Kashmira Kasat. March 19, 2019. "Introductory Chapter: Artificial Intelligence - Challenges and Applications", Accessed on: June 3, 2021. [Online]. Available: https://www.intechopen.com/books/artificial-intelligence-scope-and-limitations/introductory-chapter-artificial-intelligence-challenges-and-applications, 2019

18. Heba Soffar. March 19, 2019. "Applications of Artificial Intelligence in the space industry & NASA Artificial Intelligence", Accessed on: June 3, 2021. [Online]. Available: https://www.online-sciences.com/robotics/applications-of-artificial-intelligence-in-space-industry-nasa-artificial-intelligence

19. F. Saeed, N. Gazem, F. Mohammed and A. Busalim (Eds). 2018. Recent Trends in Data Science and Soft Computing. *Proceedings of the 3rd International Conference of Reliable Information and Communication Technology (IRICT 2018)*, Kuala Lumpur, Malaysia (Vol. 843), Springer.

20. "Goals of Artificial Intelligence", June 3, 2021. Accessed on: June 3, 2021. [Online]. Available: https://poonam3958.wordpress.com/goals-of-ai/

21. Y. Siriwardhana, G. Gür, M. Ylianttila and M. Liyanage. 2020. "The Role of 5G for Digital Healthcare Against COVID-19 Pandemic: Opportunities and Challenges", *ICT Express*, 7(2), 244–252.

22. E. Hossain and M. Hasan. 2015. "5G Cellular: Key Enabling Technologies and Research Challenges", *IEEE Instrumentation & Measurement Magazine*, 18(3), 11–21.

2 Attacks, Security Concerns, Solutions, and Market Trends for IoT

Deepak Rai and Shyam Singh Rajput
National Institute of Technology Patna, Patna, India

Rakesh Kumar
National Institute of Technology Hamirpur, Hamirpur, India

CONTENTS

2.1 INTRODUCTION

The Internet of Things (IoT) links all types of physical objects to the internet. IoT devices are integrated systems capable of detecting, working with each other and interacting with the world. These devices are also capable of data sharing and acting autonomously on events in the real world. IoT applications are widely used on virtually all mobile devices, and are wirelessly connected to the internet. By the end of 2020, the exponential growth of IoT devices had made the global IoT industry worth US$212 billion, and projections indicate that by 2025 this will hit approximately US$1.6 trillion [1]. Organizations need an intelligent mechanism which is able to automatically detect suspicious IoT devices linked to their networks. The reasons for this are the broad usage,

DOI: 10.1201/9781003045809-3

variability, standardization barriers and intrinsic flexibility of these devices. Available solutions for protecting the whole IoT network are inadequate because of resource constraints, heterogeneity, the vast amount of IoT device-generated real-time data and dynamic network behavior.

Attackers may try to access the vulnerabilities that interact directly with database backend systems in application layer protocols, such as message queue telemetry transport, hypertext transfer protocol, and the domain name system. Security breaches may result from successful manipulation of one or more of these protocols. Machine learning is seen as an alternative tool to protect against malware, botnets and other attacks. A promising approach to IoT device vulnerabilities is to embed network-level solutions where the inflow and outflow of the IoT devices is managed so that they work without any problem [2].

2.1.1 LAYERED ARCHITECTURE OF IoT NETWORK

The fundamental IoT architecture has four basic stages (Figure 2.1). The generic IoT architecture corresponding to different network layers is discussed in two parts [3]. Figure 2.2 shows various IoT devices and associated services at every layer. Figure 2.3 is a block diagram showing the sequential functioning of each network layer corresponding to the IoT.

2.1.2 BUILDING BLOCKS OF IoT SYSTEM

There are several approaches to IoT architecture, the effectiveness and applicability of which directly correlate with the quality of its building blocks and the way they interact [4]. One such building block is shown in Figure 2.4 and Table 2.1.

2.2 IoT DEVICES

IoT devices are basically intelligent devices that enable internet connectivity and are able to communicate with other devices over the internet and provide a user with remote access to control the device as necessary. The number of IoT devices today exceeds the total number of people on the

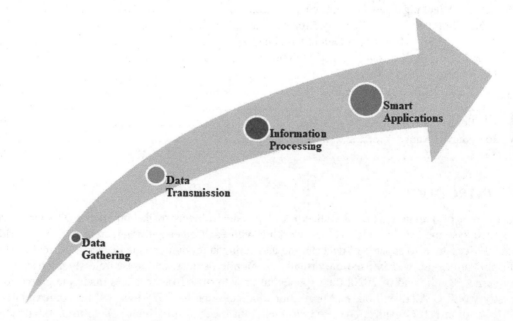

FIGURE 2.1 Fundamental architecture of IoT.

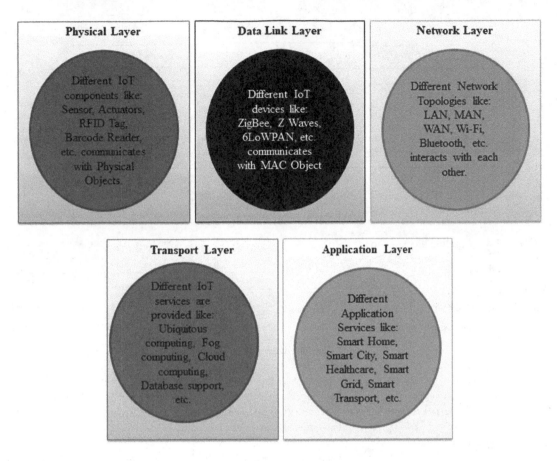

FIGURE 2.2   IoT devices and services at different network layers.

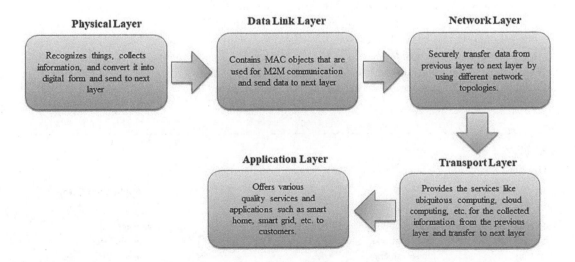

FIGURE 2.3   Functioning of each network layer corresponding to IoT.

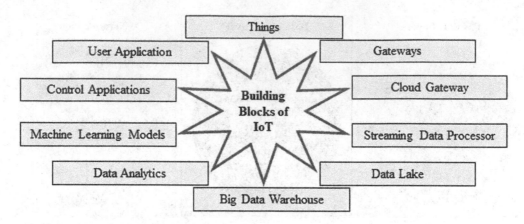

FIGURE 2.4 Building blocks of IoT.

TABLE 2.1
Building Blocks of IoT

S.No.	IoT Building Block	Description
1	Things	Object which gathers data through sensors and sends them to other network for actions through actuators
2	Gateways	Filters the data before moving it to IoT cloud from things
3	Cloud gateways	Handles the secure transmission of data between cloud and in field gateways
4	Streaming data processor	Allows data transition to data lake
5	Data lake	Stores generated data in original format
6	Big data warehouse	Contains cleaned data
7	Data analytics	Finds trends between data
8	Machine learning models	Creates more efficient model for control
9	Control applications	Sends automatic command to actuators
10	User applications	Enables user connection to IoT system

Earth. At present, there are around 7.62 billion people on Earth and it is estimated that in 2021, nearly 20 billion smart IoT devices with 5G network will be running.

The production and use of IoT devices is rising quite rapidly. IoT devices support the extension of internet connectivity outside the scope of existing devices like cell phones, computers etc. These IoT devices are equipped with the latest Wi-fi technologies that allow them to be controlled remotely and interact over the internet. Some IoT devices are shown in Table 2.2.

TABLE 2.2
IoT Devices

Name of IoT Device	Purpose	Application Category
Bitdefender BOX	IoT security solution	Home security
Google Home	Voice controller	Home appliance
Amazon Echo	Voice controller	Home appliance
Nest Cam	Indoor camera	Home security
Mr Coffee	Smart coffee maker	Home appliance

(Continued)

TABLE 2.2 (Continued)

Name of IoT Device	Purpose	Application Category
SmartMat	Intelligent yoga mat	Health and fitness
Philips Hue	Connects light and moves it around the way you like	Home energy management
TrackR bravo	Coin-sized tracking device	Home security
Linquet	Bluetooth tracking device	Home security
Amazon Echo Spot	Smart alarm clock	Home appliance
BB8 SE	Droid with force band	Information and entertainment
Nest Thermostat	Learns what temperature you like	Home energy management
Amazon Echo Plus	Voice controller	Home appliance
Logitech Pop	Smart button controller	Home appliance
Nest Cam Outdoor	Outdoor camera with advanced night vision	Home security
AWS IoT Button	Programmable dash button	Home appliance
Logitech Circle	Portable Wi-fi video camera	Home security
Logitech Harmony	Integrates control of connected lights, music, thermostat etc.	Home appliance
Awair	Smart air-quality monitor	Home appliance
Nest Protect	Smoke alarm with industrial-grade smoke sensor	Home security
Navdy	Combines high-quality projection with voice and gesture	Automative
Hydrawise	Manages irrigation controller	Home appliance
Sync Smartband	Activity tracker	Health and fitness
June	Intelligent oven	Home appliance
August	Outdoor camera	Home security
Singlecue	Gesture controller	Home appliance
Triby	Smart portable speaker	Entertainment
Kinsa	Smart thermometer	Health and fitness
Ring Pro	Smart video doorbell	Home security
Withings	Blood-pressure monitor	Health and fitness
Fitbit Surge	Smart watch for continuous heart-rate monitoring	Health and fitness
MaxMyTV	TV automation	Entertainment
Sense	Sleep tracker	Health and fitness
PulseOn	Heart-rate monitor	Health and fitness
Keen Home	Smart vent system	Home energy management
Parrot	Wireless plant monitor	Home appliance
Footbot	Indoor air-quality monitor	Home energy management
Anova	Prepares food and places it in re-sealable bag	Home appliance
Garageio	Controls garage from anywhere	Home appliance
AIRBIQUITY	Sends prompt notification in the event of vehicle crash	Connected cars
DASH	Provides alerts for present and future maintenance requirements of the vehicle	Connected cars
ZUBIE	Monitors vehicle health and driver performance	Connected cars
INSTEON	Monitors common in-home occurrences like water leaks, door lock, smoke etc.	Home appliance
PEAK 8 CONNECTED	Facilitates the measurement of soil level	Agriculture
HERDDOGG	Collects data related to herds and sends to cloud	Agriculture
PROPELLER	Discovers what triggers asthma attacks	Health and fitness
Netatmo Welcome	Indoor built-in facial recognition	Home security
Neurio	Provides information on energy-saving opportunities	Home energy and management
Awarepoint	Tracks the location of employees, assets, customers etc.	Industrial
Aptomar	Detects oils and gas spills	Industrial
RoboCV	Moves pallets and boxes from place to place	Industrial
Xerafy	Asset tracking	Industrial

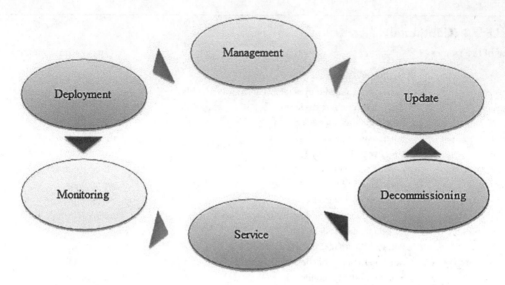

FIGURE 2.5   Life cycle of IoT devices.

2.2.1   IoT Device Lifecycle

IoT has a very simple development lifecycle (see Figure 2.5).

2.2.2   Benefits of IoT Devices

Among the many benefits of smart IoT devices are:

- Promote interaction between different IoT devices
- Provide good automation and control
- Better to operate because they are integrated with more technical information
- Capable of monitoring the features of IoT devices robustly
- Save very significant amount of time and resources
- Automate daily life tasks
- Increased efficiency.

2.2.3   Drawbacks of IoT Devices

There are also some drawbacks, including the following:

- No global acceptable compatibility standard
- Can become extremely complex, leading to failure
- Can be significantly impacted by privacy and security breaches
- Reduced user protection
- May lead to significant job reductions
- Increasing AI technology may take control of life in due course.

2.3   IoT NETWORK TECHNOLOGIES

It is clear that from the IoT applications perspective, wireless is the only way to connect such a large number of IoT devices. At the same time, several other wireless communication and networking technologies have been developed by the industry in order to fulfill the demands of IoT applications.

TABLE 2.3
IoT Network Technologies

Types of IoT Network		Technology	Range (in meters)	Topology Used
Short-range networks		Radio Frequency Identification (RFID)	0.01–100	Point to Point (P2P)
		Near Field Communication (NFC)	Up to 0.2	P2P
		Bluetooth Low Energy (BLE)	10–100	P2P, Star
		Ant	Up to 30	P2P, Star, Mesh
		EnOcean	30–300	Mesh
		Z-Wave	40–200	Mesh
		Insteon	30–50	Mesh
		ZigBee	10–100	Star, Mesh, Tree
		MiWi	10–100	Star, Mesh, Tree
		DigiMesh	10–100	P2P, Mesh
		WirelessHART (Highway Addressable Remote Transducer)	10–100	Mesh
		Thread	10–100	Star, Mesh, Tree
		6LowPAN	10–100	Star, Mesh, Tree
		Wi-Fi	Up to 100	Star
		Low Power Wi-Fi (Wi-Fi HaLow)	Up to 1000	Star
Long-range networks	Licensed	Narrow Band IoT (NB-IoT)	10000–15000	Star
		Enhanced Machine Type Communication (eMTC)	10000–15000	Star
		Extended Coverage GSM IoT (EC-GSM-IoT)	10000–15000	Star
	Unlicensed	Long-Range WAN (LoRaWAN)	10000–15000	Star of Stars
		Symphony Link	10000–15000	Star
		Weightless	2000–5000	Star
		SIGFOX	10000–50000	Star
		DASH7	2000–5000	Star or Tree

On the basis of area under coverage, network technologies can be classified into two groups: short range and long range [5]. Long-range networks can be either licensed or unlicensed. Some of the IoT network technologies are shown in Table 2.3.

2.4 DATA AGGREGATION IN IoT

Data collection and analysis requirements present the greatest challenge nowadays because of the substantial increase in the number of smart objects and their applications. Data aggregation is a mandatory task in IoT, as the heterogeneous data from various sources lead to higher energy requirements. In this case, one of the energy-saving solutions is the processing and compilation of data before summarization and submission. Data aggregation techniques are categorized as shown in Figure 2.6. The protocols and algorithms used for data aggregation are shown in Figure 2.7.

2.5 ATTACKS AND SECURITY THREATS IN IoT

From an economic point of view, security and privacy are the two main considerations with IoT services and applications, and these have been thoroughly studied from different viewpoints [6]. The current internet platform allows almost all types of security attacks, ranging from basic hacks to well-coordinated security breaches. This adversely affects different industries. The limitations of IoT devices cause a significant increase in security constraints for both applications and devices [7].

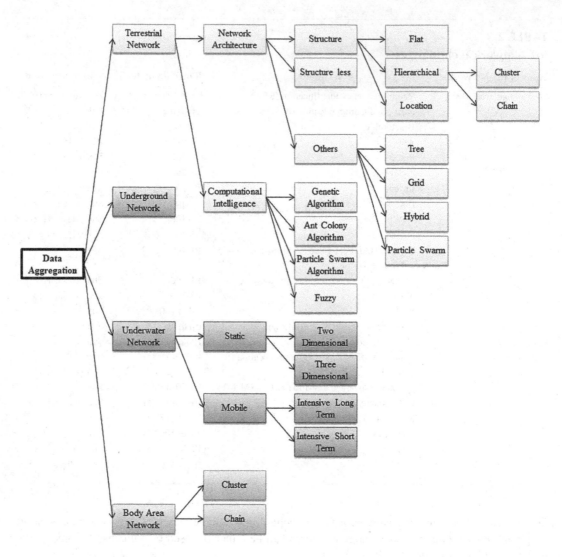

FIGURE 2.6 Data aggregation techniques.

If the IoT is to be effectively realized, it is crucial to explore the origins of security and privacy problems, and in particular, it is important to know to know whether the IoT protection issues are new or a renewal of the legacy of old technologies. Here, we give an overview of IoT security, attacks and their solutions.

2.5.1 ATTACKS ON DIFFERENT NETWORK LAYERS

Attacks in IoT can be abstractly categorized as physical (Table 2.4), data link, (Table 2.5) network (Table 2.6), transport (Table 2.7) and application layer (Table 2.8) attacks.

2.5.2 ATTACKING TOOLS USED IN IoT

This section lists existing tools used by IoT device attackers. A wide range of such tools is available in the market. Two categories of tools are discussed: hacking tools (Table 2.9) and sniffing tools (Figure 2.8).

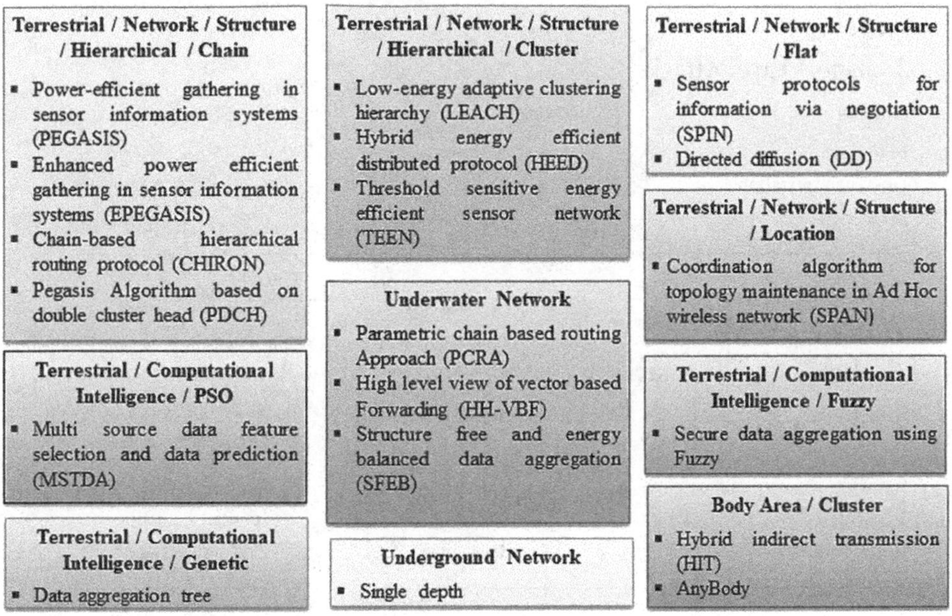

Terrestrial / Network / Structure / Hierarchical / Chain	Terrestrial / Network / Structure / Hierarchical / Cluster	Terrestrial / Network / Structure / Flat
▪ Power-efficient gathering in sensor information systems (PEGASIS) ▪ Enhanced power efficient gathering in sensor information systems (EPEGASIS) ▪ Chain-based hierarchical routing protocol (CHIRON) ▪ Pegasis Algorithm based on double cluster head (PDCH)	▪ Low-energy adaptive clustering hierarchy (LEACH) ▪ Hybrid energy efficient distributed protocol (HEED) ▪ Threshold sensitive energy efficient sensor network (TEEN)	▪ Sensor protocols for information via negotiation (SPIN) ▪ Directed diffusion (DD)
		Terrestrial / Network / Structure / Location ▪ Coordination algorithm for topology maintenance in Ad Hoc wireless network (SPAN)
Terrestrial / Computational Intelligence / PSO ▪ Multi source data feature selection and data prediction (MSTDA)	**Underwater Network** ▪ Parametric chain based routing Approach (PCRA) ▪ High level view of vector based Forwarding (HH-VBF) ▪ Structure free and energy balanced data aggregation (SFEB)	**Terrestrial / Computational Intelligence / Fuzzy** ▪ Secure data aggregation using Fuzzy
Terrestrial / Computational Intelligence / Genetic ▪ Data aggregation tree	**Underground Network** ▪ Single depth	**Body Area / Cluster** ▪ Hybrid indirect transmission (HIT) ▪ AnyBody

FIGURE 2.7 Data aggregation protocols and algorithms.

TABLE 2.4
Physical Layer Attacks

Attacks	Detected or Prevented	Source
Hardware Trojan	Detected	Trojan activation
Physical tampering	Detected	Circuit modification
Denial of services	Detected	Personal firewall
Eavesdropping	Detected	Kill/sleep command

TABLE 2.5
Data Link Layer Attacks

Attacks	Detected or Prevented	Source
Collision/jamming	Prevented	Firewall update
Exhaustion	Prevented	Cryptographic schemes
Unfairness	Prevented	Short packet frame

TABLE 2.6
Network Layer Attacks

Attacks	Detected or Prevented	Source
Routing based	Detected	Reliable routing
Sybil	Prevented	Validation of identities
Black hole	Detected	Reliable routing
Sleep deprivation	Detected	Firewall update

TABLE 2.7
Transport Layer Attacks

Attacks	Detected or Prevented	Source
Flooding	Prevented	Client puzzle
De-synchronization	Prevented	Authentication
Spoofing	Prevented	Device ID

TABLE 2.8
Application Layer Attacks

Attacks	Detected or Prevented	Source
Malicious code injection	Detected	Preliminary test
Software-based modification	Prevented	Secure software update
Integrity	Detected	Outlier detection
Brute force	Prevented	Strong password

TABLE 2.9
Hacking Tools

S.No.	Hacking Tool	Attack Type
1	Low Orbit Ion Canon (LOIC) tool	DDoS
2	HTTP Unbearable Load King (HULK) tool	DDoS
3	DDOS simulator tool	DDoS
4	GoldenEye HTTP denial of service tool	DDoS
5	Slowloris tool	DDoS
6	R-U-Dead-Yet (RUDY) tool	DDoS
7	Hping tool	DDoS
8	Ettercap tool	Eavesdropping and traffic analysis
9	SSLStrip	Eavesdropping and traffic analysis
10	Evilgrade	Eavesdropping and traffic analysis
11	Aircrack-NG	Eavesdropping and traffic analysis
12	SQLmap	Malware and code injection
13	oclHashcat	Dictionary, Hybrid, Mask, and Rule Base
14	Ncrack	Brute force
15	Cain	Brute force
16	Abel	Brute force
17	Tor Network	Network surveillance
18	Zmap	Security scanner
19	Cisco SNMP-Slap	DNS spoofing
20	Smikims-arpspoof	ARP spoofing
21	RouterSploit	Router exploitation framework
22	HackRF	Replay attack
23	Cross-site scripting	Attacks on web applications
24	SQL injection	Computer Security Vulnerability
25	Rainbow Crack	Password cracking
26	Log Injection-Tampering-Forging	Attack on log files
27	NetCat	Port scanning
28	Wireshark snooping	Packet analyzer

(Continued)

TABLE 2.9 (Continued)

S.No.	Hacking Tool	Attack Type
29	Net Start Netlogon	Prevent to access network
30	OTG-Autz-003	Privilege escalation
31	Jailbreak tool	Hack iOS devices
32	Trojan-Netbus	Key injection
33	Spyware Cool	Keylogging, Trojan horse
34	VirTool	Spyware
35	Rootkits	Malware

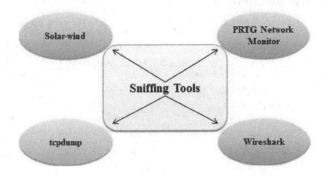

FIGURE 2.8　Sniffing tools.

2.5.3　Solutions to IoT Security Attacks

There is significant existing work describing security attack mitigation techniques for the IoT network.

2.5.3.1　Traditional Defense Techniques

- IP spoofing attacks, DDoS attacks, fragment attacks, routing attacks, and so on are avoided by standard security methods such as packet filtering using firewalls and proxies.
- The IoT network is also covered by IDS (intrusion detection system) and IPS (intrusion prevention system).
- IDS is based on anomalies, signatures and hybrid.
- By matching payload/heads with written rules, IDS detects threats and cuts the link from the source IP address.
- IPS provides an immediate response if threats are detected by matching payloads/heads with written rules, preventing malicious activities over the network.

2.5.3.2　Current Defense Techniques

- In IDS, artificial intelligence (AI) methods are also used to avoid IoT attacks. AI applications can detect threats competently and automatically update their database to avoid zero-day attacks, DoS attacks, injection of malicious software, and so on.
- The blockchain technique can also be used in IoT network traffic for data integrity, protection and privacy, shielding IoT devices from various vulnerabilities [8]. By checking the digital signatures on the blockchain network, blockchain technology authenticates IoT devices [9].
- Fog computing also provides a wide range of advantages, such as enhanced security, and reduced bandwidth and latency, and solves the cloud computing storage dilemma [10].
- It is also possible to see machine learning as an alternative tool for defense against ransomware, botnets and other attacks [11]. A promising approach to IoT device vulnerabilities is to

embed network-level solutions that track network traffic to/from IoT devices to ensure that they run normally and identify suspicious behaviors [12]. By deploying a deep neural network model that will train the model for distinct network data such as payload, no packets, SMTP and packet arrival time, we can classify the IOT traffic into attack and benign traffic. Such data can be derived from a real-time scenario and can be checked to detect whether the IOT network is malicious or invasive.

Table 2.10 shows possible solutions to some attacks.

TABLE 2.10
IoT Security Attacks and Solutions

Security Threat Category	Attacks	Solution
DDoS/DoS (denial of service)	Collision attack	Personal firewall, IDS, NIDS, and cryptographic schemes, etc.
	Channel congestion attack	
	Battery exhaustion attack	
	PAN Id conflict	
	Fragmentation attack	
	Flooding attack	
Eavesdropping and traffic analysis	Sniffing attack	Use of Kill or Sleep command, blocking, cryptographic schemes etc.
	Snooping attack	
	MITM attack	
Spoofing	IP spoofing	Using device Id protocols IPSec, TLS, and SSH etc.
	DNS spoofing	
	ARP spoofing	
	DDoS spoofing	
	Hello Flood	
Routing	Sinkhole attack	Using reliable routing, RPL, Watchdog, reputation, and trust strategies.
	Selective forward attack	
	Black hole attack	
	Wormhole attack	
	Replay attack	
Side channel	Cache attack	Using circuit or design modification, isolation, and blocking etc.
	Timing attack	
	Power monitoring attack	
	Electromagnetic attack	
	Fault attack	
Malware attack	Malicious code attack	Using preliminary test, software integrity, outliers detection etc.
	Malicious data injection	
Repudiation attack	Repudiation attack	Using cryptographic network protocols.
Information disclosure attack	Snooping attack	Using cryptographic authentication schemes.
User privacy	Spyware	Using anti-virus, Trojan activation, side channel analysis, etc.
	Trojan	
	Sniffer	
	Virus	
	Malicious codes	

2.6   OPTIMIZATION OF IoT NETWORK

Although the IoT has evolved tremendously worldwide over the past decade, there remain several gray areas where it can be improved. IoT devices consume an enormous amount of power because of the use of wireless sensors for continuous data collection. This can significantly decrease the sustainability of the IoT network and the lifetime of the IoT devices. It is therefore imperative to optimize the energy consumption of IoT devices.

There are several metaheuristic and evolutionary optimization models for saving energy. In this chapter we present one metaheuristic-based optimization algorithm using a hybrid algorithm with a combination of the Whale Optimization Algorithm (WOA) and the simulated annealing (SA) method. The use of SA in the proposed algorithm helps enhance the WOA search process. The four different fitness functions used in this problem are shown in Figure 2.9.

2.6.1   WOA Algorithm

One of the metaheuristic algorithms inspired by nature is WOA, which is based upon the hunting habits of humpback whales. The attacking pattern of whales against prey is called bubble-net feeding, which is carried out when bubbles are produced around the prey in a circle formation. This mechanism is composed of three phases shown in Figure 2.10.

2.6.2   Simulated Annealing

Simulated annealing (SA), an optimization technique based on the principles of thermodynamics, is mainly used for obtaining the most appropriate global minimum from the group of several local minimums. It uses an objective function to find the global minimum. It is quite similar to the hill-climbing search technique used in artificial intelligence, the only difference being that in hill climbing, the best move is always selected, while in SA a random move is selected. If the probability of a random move being selected is less than 1, then the selected move is abandoned and another move is selected.

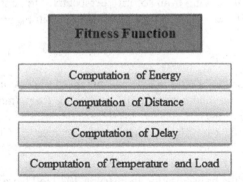

FIGURE 2.9   Fitness functions for optimization.

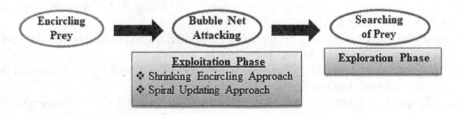

FIGURE 2.10   Phases of WOA algorithm.

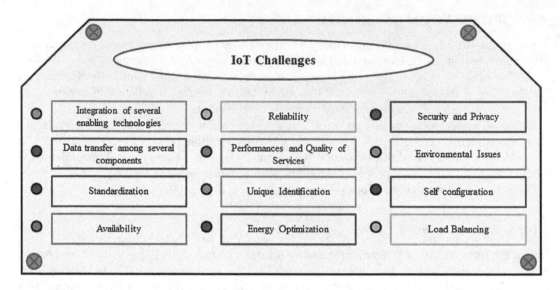

FIGURE 2.11   Challenges in IoT.

In WOA-SA, as an initial state, the SA ensures the replacement of a blind operator by a local search consisting of the solution itself. The improved solution later replaces the initial solution after operating on the local search. The SA method helps to boost the WOA's efficiency, improving its power of exploitation, which helps in finding the optimal solution.

2.7   CHALLENGES IN IoT

The IoT is becoming part of all our lives, but there are still several challenges to its worldwide adoption, growth and development, ranging from security perspectives to people's expectations[13, 14]. Some of the difficulties of efficient IoT implementation are seen in Figure 2.11.

2.8   IoT MARKET ANALYSIS

The IoT is at the heart of a thriving sector in an environment of hyper-connectivity in which countless innovators provide software and hardware for everything from smart homes and vehicles to medical devices and manufacturing.

Among the industries hardest hit by the rough economic reset of COVID-19 are automotive, retail, wholesale trade, and transport.

- IoT sales estimates for the transportation industry in 2020 fell from $43 billion to $34 billion, according to an estimate by Gartner.
- IoT for enterprise and automotive fell to $351 billion from $393 billion.
- Revenue estimates from IoT healthcare providers are among the most positive, expected to grow from $25 billion to $29 billion.
- In just a year, there has been a 27 percent rise in IoT startups and related companies on Crunch base, rising from 26,792 in 2019 to 34,120 today.
- IoT startup funding achieved a 15 percent year-on-year rise from 2019 to 2020, reaching $4.7 billion, according to Venture Scanners IoT Funding Update.
- According to Technology Business Research's latest forecast, the worldwide commercial IoT market will grow from $385 billion in 2019 to $687 billion in 2025, attaining a compound annual growth rate (CAGR) of 10.1%.

TABLE 2.11
Top IoT Startups in 2021

IoT Company	Category	Primary Service	Location
RingCentral	Cloud computing, communication, and mobile	Cloud-based communication	Belmont, California
Jiobit	Children and family	Location tracker	Chicago, Illinois
Cooler Screens	Adtech	Real-time promotions	Chicago, Illinois
CCC Information Services	Automotive and insurance	Hyper-scale technologies and apps	Chicago, Illinois
Farmer's Fridge	Food	Handcrafted food in jars	Chicago, Illinois
AlertMedia	Telecommunications and healthcare	Mass alert communication	Austin, Texas
SimpliSafe	Security	Wireless/cellular home security systems	Boston, Massachusetts
Superpedestrian	Robotics and transportation	Human-powered mobility	Cambridge, Massachusetts
10th Magnitude	Healthcare and travel	Cloud-based solution for communication	Chicago, Illinois
CSG	Wireless and finance	BSS software service	Greenwood Village, Colorado
Inspire	Cleantech	Cleaner and efficient power	Culver City, California
Rogue Waves	Healthcare, finance and entertainment	API management	Louisville, Colorado
Enevo	Food, retail, and cleantech	Waste and recycling	Boston, Massachusetts
Nexleaf Analytics	Big data	Lightweight sensor	Los Angeles, California
Temboo	Software	Internet-enabled automated systems	New York
Wink	Smart homes	Light, power and security systems in homes	New York
Point Inside	Retail and healthcare	Data-driven positioning	Bellevue, Washington
Armis	Security	Comprehensive device and asset recovery	Palo Alto, California
Atomiton	Energy and agriculture	Secure interconnection and interaction	Sunnyvale, California
Flutura	Energy and AI	Asset uptime and operation efficiency	Houston, Texas
FogHorn Systems	Robotics and industrial	Edge intelligence	Mountain View, California
MagicCubes	Finance and automotive	Trusted execution environment and security platform	Sunnyvale, California
Tive	Software	Supply chain	Cambridge, Massachusetts
Xage Security	Energy and transportation	Blockchain technology-based security platform	Palo Alto, California
Avaamo	Insurance, finance and healthcare	IoT services	Los Altos, California
Sigmaways	Finance and gaming	Churn and loyalty analysis	Fremont, California
Nodle.Io	Mobile	Low energy-powered network	San Francisco, California
Samsara	Transportation and food	Sensor data solutions	San Francisco, California
Eero	Consumer electronics	Reliable internet access	San Francisco, California
ScienceSoft	Miscellaneous	IoT application development and data analytics	McKinney, Texas
Oxagile	Miscellaneous	IoT hardware prototyping and integration	New York
R-Style Lab	Miscellaneous	IoT mobile and web	San Francisco, California
HQ Software	Miscellaneous	High-level development of device and sensors	Boston, Massachusetts
PTC	Miscellaneous	Deployment of connected solutions	Boston, Massachusetts

While IoT services are also provided by tech giants such as Google, Intel, Microsoft or Amazon, here we will concentrate on medium-sized organizations and startups with unique IoT products. Table 2.11 lists some of the leading IoT startups to watch in 2021.

REFERENCES

1. Statista, Size of the internet of things (IoT) market worldwide from 2017 to 2025, accessed on: 2021-01-05. [Online]. Available: https://www.statista.com/statistics/976313/global-iot-market-size/
2. M. A. Khan and K. Salah, IoT security: Review, blockchain solutions, and open challenges, *Future Generation Computer Systems*, vol. 82, pp. 395–411, 2018.
3. Hichem Mrabet, Sana Belguith, Adeeb Alhomoud, and Abderrazak Jemai. A survey of IoT security based on a layered architecture of sensing and data analysis, *Sensors*, vol. 20, no. 13, p. 3625, 2020.
4. V. Adat and B. Gupta. Security in internet of things: Issues, challenges, taxonomy, and architecture, *Telecommunication Systems*, vol. 67, pp. 423–441, 2018.
5. Jozef Mocnej, Adrian Pekar, Winston K. G. Seah, and Iveta Zolotova. *Network Traffic Characteristics of the IoT Application Use Cases*, School of Engineering and Computer Science, Victoria University of Wellington, 2018.
6. Ranjit Patnaik, Neelamadhab Padhy, and K. Srujan Raju. A systematic survey on IoT security issues, vulnerability and open challenges, In Satapathy, S., Bhateja, V., Janakiramaiah, B., Chen, Y.W. (eds) *Intelligent System Design*, pp. 723–730. Springer, Singapore, 2021.
7. M. Abolhasan Makhdoom, J. Lipman, R. P. Liu, and W. Ni, Anatomy of threats to the internet of things, *IEEE Communications Surveys & Tutorials*, vol. 21, pp. 1636–1675, 2018.
8. Noshina Tariq, Muhammad Asim, Farrukh Aslam Khan, Thar Baker, Umair Khalid, and Abdelouahid Derhab. A blockchain-based multi-mobile code-driven trust mechanism for detecting internal attacks in internet of things, *Sensors*, vol. 21, no. 1, p. 23, 2021.
9. W. Ejaz and A. Anpalagan, Blockchain technology for security and privacy in internet of things. *Internet of Things for Smart Cities*, pp. 47–55, 2019.
10. A. Mutlag, M. K. A. Ghani, N. A. Arunkumar, M. A. Mohamed, and O. Mohd, Enabling technologies for fog computing in healthcare IoT systems, *Future Generation Computer Systems*, vol. 90, pp. 62–78, 2019.
11. Hussain, Fatima, Rasheed Hussain, Syed Ali Hassan, and Ekram Hossain. Machine learning in IoT security: Current solutions and future challenges, *IEEE Communications Surveys & Tutorials*, vol. 22, pp. 1686–1721, 2020.
12. A. Kumar and T. J. Lim, Edima: Early detection of IoT malware network activity using machine learning techniques, In *2019 IEEE 5th World Forum on Internet of Things (WF-IoT)*, pp. 289–294, 2019.
13. Dhuha Khalid Alferidah, and N. Z. Jhanjhi. A review on security and privacy issues and challenges in internet of things, *International Journal of Computer Science and Network Security IJCSNS*, vol. 20, no. 4, pp. 263–286, 2020.
14. Jayashree Mohanty, Sushree Mishra, Sibani Patra, Bibudhendu Pati, and Chhabi Rani Panigrahi. IoT security, challenges, and solutions: A review, *Progress in Advanced Computing and Intelligent Engineering*, pp. 493–504, 2021.

3 Intelligence and Security in the 5G-Oriented IoT

Jasminder Kaur Sandhu and Prateek Srivastava
Chitkara University Institute of Engineering and Technology,
Chitkara University, Punjab, India

Yadunath Pathak
Indian Institute of Information Technology (IIIT-Bhopal), Bhopal, India

Meena Pundir
Chitkara University Institute of Engineering and Technology,
Chitkara University, Punjab, India

CONTENTS

3.1   INTRODUCTION

The 5G-oriented Internet of Things (IoT) has created much interest and is a popular buzzword for technological advancements in various industries. 5G is creating a new standard for global wireless technology. It can reach a download speed of 20 GB/s. The new 5G technology has been designed to virtually connect multiple components: devices, instruments, machines, the cloud. It delivers very high speed, low latency, power efficiency, huge network capacity, better availability, and reliability, resulting in high-quality human experience [1]. While no one has ownership of this technology, multiple mobile and telecommunications industry companies are contributing to its development and implementation. Qualcomm has played a significant role in developing the foundation technologies of 5G [2].

5G works by modulating a digital signal across multiple channels, which is also known as Orthogonal Frequency Division Multiplexing (OFDM). 5G also uses a 5G New Radio (5G NR) air interface. Technologies such as sub-6 GHz and mmWave have been used for achieving wider bandwidth in 5G.

IoT technology enables interconnection of devices that will benefit from next-generation connectivity. Any object or system can be assigned an Internet Protocol (IP) to transfer data over the network [3]. Different types of devices, mechanical equipment, machines, and objects are assigned Unique IDentifiers (UIDs) to transfer the data. IoT devices help gather and monitor data continuously, optimizing processes and operations. IoT hardware enables the transfer of data without human–computer interaction. IoT is a connected ecosystem for smart devices, and consists of an embedded system (internet connectivity, hardware, sensor with communication and control features). Human intervention is only required at the time of installation of IoT devices, not during their operation [4]. Terahertz band communication is used to manage spectrum and antenna design.

3.1.1   5G INTEGRATION WITH IoT

The 5G network will transform society, help improve quality of life and ease of doing business. With the 5G infrastructure, a large amount of data is generated at a very high speed and can be simultaneously optimized to improve services, industrial needs. Integration of the 5G network with IoT is also called Machine-to-Machine (M2M) communication [5].

5G-integrated IoT applications are available for mobile IoT variants of smart cities, improving logistics and smart energy metering. Industries that will benefit from the 5G IoT include [6, 7]:

- Automotive and transportation
- Smart grid
- Smart factories
- Energy-efficient buildings
- Smart cities
- Utilities
- Security and surveillance
- Agriculture
- Retail
- Healthcare
- Aerospace

Figure 3.1 illustrates the IoT paradigm integrated with 5G technology. IoT enables the interconnection of various devices and 5G technology renders high-speed communication throughout the network.

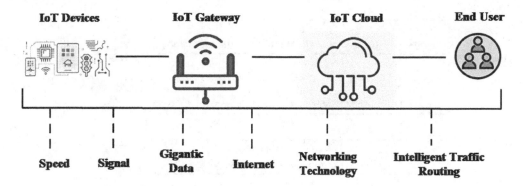

FIGURE 3.1 The integration of IoT and 5G.

This chapter introduces a novel perspective on the integration of IoT with the help of 5G technology. Section 2 discusses the detailed components of the IoT–5G integration, and the layered structure of the integration. Section 3 illustrates the functional and non-functional characteristics of the interconnected system. Section 4 elaborates the security aspect based on the layered structure. Section 5 illustrates the concept of intelligence in this framework. The tools are presented in Section 6. The conclusion summarizes the issues and suggests future research directions.

3.2 DETAILED COMPONENTS OF IoT–5G INTEGRATION

The IoT includes the following components:

- *Sensors.* Every IoT device is developed for implementation of a particular task. It consists of an embedded system having single or multiple sensors to measure single and multivariable data. Sensors are used to acquire data from different sources using IoT devices.
- *Data.* All the data received from IoT devices are transferred through the 5G network ecosystem. Data can be in different forms such as numbers, text, images, and videos.
- *Analytic engines.* All the data is transferred from the IoT device to the cloud, where it is used to further improve or optimize the system. For optimization of the system or process, analytics and human experience-based intervention or machine-learning algorithms are required.

The main components [7, 8] of 5G technology are:

- *Local server.* The data collected from local workstations are stored on the local server. This local server operates in a limited region of interest.
- *Central server.* The information collected at local servers is then processed to remove all the erroneous information and unnecessary repetition. Processed data are accumulated at the central server for main processing and decisions. This makes the component architecture a completely distributed system for information collection and processing.
- *5G small cell.* Small-cell technology is used by the telecommunications companies to roll out the 5G network. It consumes low power and uses short-range wireless transmission to cover small distant spaces. The features of a small cell are similar to those of base stations in the telecommunications industry.
- *5G macro.* Multiple Input, Multiple Output (MIMO) antennas with multiple elements are used by 5G macro cells with connections to simultaneously send or receive data. High throughput can be maintained even if more people are connected.

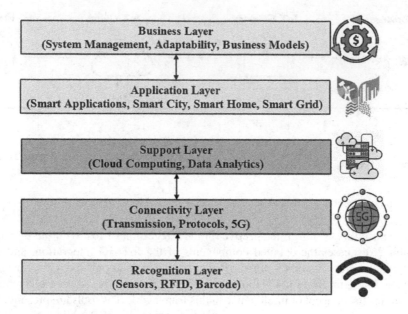

FIGURE 3.2   Layered architecture of IoT–5G integration.

3.2.1 LAYERED ARCHITECTURE

This section describes the layered architecture which enables the various IoT-based application areas to perform their tasks much more rapidly and accurately using 5G technology [9, 10]. Figure 3.2 shows the five layers of integration.

Recognition layer. Sensor data must be protected from unintended access, key agreement for data transmission only to authorized persons, protection of physical devices. This layer deals with both hardware and software security.

Connectivity layer. Integrity, encryption, authentication and optimized protocols are required to speed up the data transmission process. This paves the way for the integration of 5G technology. This layer is responsible for security of the entire network.

Support layer. Proper antivirus protection, cloud security, and optimized encryption, if required by the application, are provided by this layer. The huge volume of data collected in this process is stored at the cloud servers or the data centers. This layer looks after the security aspect while storing the data.

Application layer. Privacy, application design, policies and standards acceptable worldwide form the basis of this layer, which also looks after the security of different connected systems and the service management to be provided to the users.

Business layer. This layer protects privacy and allows flow of information in a hierarchical model in accordance with the company's policies. It also looks after maintenance and adaptability, and ensures that no data are leaked outside a particular organization.

3.3 PROPERTIES

Networked system requirements can all be categorized as functional or non-functional requirements. Those entities that are quantifiable, such as reliability, are called functional requirements [11]. Those that cannot be quantified, such as security and intelligence, are known as non-functional.

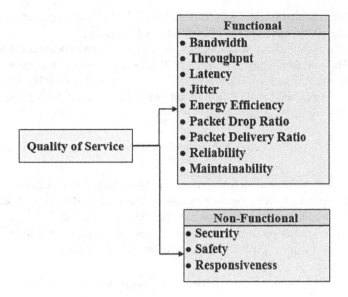

FIGURE 3.3   Quality of service.

3.3.1   QUALITY OF SERVICE

Quality of Service (QoS) is a significant feature of the 5G-oriented IoT. It is responsible for the improvement of channel bonding used for high data transmission rates. This significant feature helps the newly evolving 5G technology to act as a reliable and efficient communication medium between smart cities using the IoT paradigm [12]. It can be divided into two categories: functional and non-functional QoS [13, 14] (Figure 3.3).

3.3.2   FUNCTIONAL REQUIREMENTS

The QoS parameters that are quantitative or measurable are known as functional QoS: jitter, bandwidth, throughput, delay, energy efficiency, packet drop.

- *Bandwidth*. Bandwidth is defined as the total volume of data transferred over the internet connection in a specific time duration. The formula for bandwidth is:

$$Bandwidth = Capacity\ of\ the\ network\ communication\ link\ to\ transmit\ maximum\ data$$

  The unit of bandwidth is bits per second. High bandwidth increases the capacity of the 5G-oriented IoT network. It is directly correlated with the performance of the entire network.
- *Throughput*. Throughput is defined as the total count of packets transferred from sender to receiver within a specific time duration. The formula for throughput is:

$$Throughput = \frac{Total\ count\ of\ packets\ tranfered\ from\ sender\ to\ receiver}{specific\ time\ duration}$$

  The unit of throughput is bits per second. This functional parameter is very important for 5G-oriented IoT as it governs the functioning of the entire network. The data rate in 5G communication should be high and this is accomplished by the throughput parameter. Hence, it enhances the performance of the 5G-oriented IoT network.

- *Latency.* The total time delay in sending the data from source to destination and back again in a network. The unit of latency is milliseconds. Latency is counted in terms of Time To First Byte (TTFB) and Round Trip Time (RTT). TTFB is the total time taken to send the information from receiver to sender. RTT is the total time taken to send the information from sender to receiver. The formula for latency is:

$$Latency = Total\ time\ to\ send\ the\ data\ from\ source\ to\ destination\ and\ back\ again$$

Latency is important for the performance of the 5G-oriented IoT. It is inversely proportional to network performance. When latency is low, the network performs well but when it is high, the performance of the 5G-oriented IoT network deteriorates.

- *Jitter.* Jitter is defined as variation in the delay when data is transferred from source to destination. The unit of the jitter is a millisecond. The formula for jitter is:

$$Jitter = Variation\ in\ delay\ during\ data\ transmission$$

There are various reasons for jitter in the 5G-oriented IoT network, including congestion, configuration error, and improper queuing. High jitter leads to poor network performance and delayed packet delivery across the network. This leads to variation in proper sequencing of packets and delivery of messages in unusable form.

- *Energy efficiency.* The process of saving energy in an efficient manner to perform a particular task is known as energy efficiency. It is directly related to the cost. If the resources consume less energy, then the cost is automatically minimized. The formula for energy efficiency is:

$$Energy\ efficiency = \frac{Energy_{useful}}{Energy_{initial}} \times 100,$$

where $Energy_{useful}$ is the output energy and $Energy_{initial}$ is the input or initial energy during a process.

- *Packet drop ratio.* Packet drop ratio is defined as the ratio of the number of packets that fail to reach their specific destination to the total count of the packets sent to the destination. These packets are dropped across the entire network while data is communicated from source to destination. The formula for packet drop ratio is:

$$Packet\ drop\ ratio = \frac{Packets\ that\ fail\ to\ reach\ their\ destination}{Total\ number\ of\ packets\ sent\ from\ source\ to\ destination}$$

There are several reasons for packet drop such as congestion and presence of outliers in the network data. The packet drop ratio is also known as packet loss. Packet drop is inversely proportional to network performance. When packet drop is minimized, the networks function efficiently. But when it is high, the capacity of the network is diminished.

- *Packet delivery ratio.* Packet delivery ratio is defined as the ratio of the total count of packets successfully delivered at a specific destination to the total count of packets sent to the destination. The formula for packet delivery ratio is:

$$Packet\ delivery\ ratio = \frac{Packets\ successfully\ delivered\ to\ their\ destination}{Total\ count\ of\ packets\ sent\ to\ source\ to\ destination}$$

It is directly proportional to the network performance. When the packet delivery is maximized, network performance is enhanced but when it is low the performance of the network diminishes.

- *Reliability*. Reliability is a functional QoS parameter based on the quality of the data. It is important in the 5G-oriented IoT network. Reliability is defined as the capacity of the network to perform its essential task in a specific condition and for a particular time duration. The formula for reliability is:

$$R(t) = e^{-\int z(t)d(t)}$$

where R(t) is a reliability function and z(t) is the failure rate function.

- *Maintainability*. Maintainability is defined as the probability of performing a successful repair action within a given time interval. The formula for maintainability is:

$$M(t) = \int_0^t g(t)d(t)$$

where M(t) is the maintainability function and g(t) is the repair rate function.

3.3.3 NON-FUNCTIONAL REQUIREMENTS

QoS parameters that are qualitative and non-measurable are known as non-functional QoS. They include security, safety and responsiveness.

- *Security* is a mechanism to design activity in a 5G-oriented IoT network to protect the data and network usability from unauthorized users. It maintains the integrity, confidentiality and availability of the data across the network.
- *Safety* is defined as a state that helps keep people safe from threats posed by the network. 5G technology has no adverse health effects on the human body.
- *Responsiveness* is a core parameter in a 5G-oriented IoT network. It is defined as the rate at which the system completes its task with quick positive action within a specified time. It provides two-way communication across the network.

3.4   SECURITY

This section emphasizes the security and anonymity requirements for the applications of 5G-oriented IoT [15, 16]. The following are the main security requirements for this integration:

Confidentiality is a significant feature in the security of the 5G-oriented IoT. It prevents sensitive information from unauthorized disclosure and provides access to only authorized users. The users can be applications, systems, and human beings. Several kinds of attack affect the confidentiality of the information, such as stealing passwords, phishing, network traffic, routing attacks, and DDoS attacks. Sometimes confidentiality breaches occur unintentionally or accidentally, for example, publishing confidential information to web servers publicly, leaving private information in unattended systems and sending an email to the wrong recipient. Different methods ensure confidentiality in 5G-oriented IoT. Resources can be protected by using passwords for software. Encryption-based solutions enhance confidentiality for security. Training and policy are the administrative solutions used to provide extra confidentiality. Information sensitivity determines the level of confidentiality, which is directly proportional to the sensitivity. The more sensitive the information, the greater the confidentiality requirement in the network. Memory protection and cryptography are other solutions for ensuring confidentiality. When confidentiality is ensured, complexity of computation and latency increases.

Privacy of information is very important in a 5G-oriented IoT network. Data privacy can be attacked by malware in smart devices. This attack hijacks the cloud servers and affects the security of the cloud. Cloud security can be achieved through provision of encryption techniques together with data privacy solutions.

Integrity plays an important role in terms of security in 5G-oriented IoT. Integrity guarantees that no alteration, addition, or deletion to the information occurs. A cyclic redundancy check (CRC) mechanism provides integrity of the network. When data is transmitted from source to destination, error bits relating to unintentional alteration are easily identified by CRC. The checksum method for intentional alteration of information helps to verify the integrity of the information.

Authenticity is a significant feature of network security that ensures the source and destination nodes are genuine. It is easy to inject cloud malware attacks in the IoT paradigm with 5G technology, causing network communication to be disturbed and degraded. The use of codes such as information authentication codes protects data from unauthorized users.

Authorization is an important property of security which assign specific access rights to different types of clients in the network. A network administrator has different access rights from any client. Access levels are related to network resources such as files, services, computer programs and system applications.

Availability is a security feature that provides network guarantees. Authorized users can access resources such as information, files, system programs, and applications when required. Data unavailability can be caused by various factors such as hardware failure, network bandwidth and software downtime. A malicious attack such as DoS can cause data unavailability in the network when data is required. Intrusion detection and prevention mechanisms help maintain data availability in the 5G-oriented IoT network. Different paths are provided to reduce failover in the network when data is transmitted from source to destination.

Time criticality is a significant feature used in a real-time network. It is the length of time within which a specific task must be completed in a real-time environment. The other name for time criticality is deadline. Time-sensitive data present in the network is to be transmitted to its destination by a specified deadline.

Trust is a primary element of security in a 5G-oriented IoT network. Sometimes abnormal activity occurs in the network due to the presence of outliers. A combination of direct trust and indirect trust helps to maintain the integrity of the network and provides user privacy. The third party assigns the trust value to the service provider according to its estimated value.

Intrusion detection (ID) and predication mechanisms are an essential part of security in a 5G-oriented IoT network. Detection systems can be categorized as: anomaly detection, misuse detection and specific detection. While various ID techniques have traditionally been used for security, the current IoT 5G technology paradigm requires lightweight ID mechanisms to identify anomalies present in the network. Early prediction of any attack will improve network usage. A fingerprint-based hybrid biometric method is used to detect traditional attacks in 5G-oriented IoT network devices.

Non-repudiation is an important feature of network security in 5G-oriented IoT. It is defined as a legal service that provides proof of the source and integrity of the data. It provides secure communication, ensuring that nobody can deny the authenticity of the signature used to send information across the network. As already described, the 5G-oriented IoT architecture comprises five layers: recognition layer, connectivity layer, support layer, application layer and business layer [17, 18]. Different types of attack on a specific layer and the security action required to protect data from malicious activity are shown in Table 3.1.

TABLE 3.1

Layer-Wise Security Action, Attack, and Deliverables in a 5G-Oriented IoT Network

Layers of 5G in IoT	Layer-wise Security Action	Intrusion	Deliverables
Recognition layer	• Sensor data must be protected from unintended access • Key agreement for data transmittal to only authorized personals • Protection of physical devices	• False information • Timing • Jamming • Tampering	• Hardware security • Software security
Connectivity layer	• Integrity • Encryption techniques • Optimized protocols required for communication	• MITM • Spoofing • Eavesdropping • Sinkhole • Sybil	• Network security
Support layer	• Proper antivirus protection • Cloud security • Optimized encryption	• Malicious insiders • Virus • Account Hijacking • Cloud malware Injection attack • Hacking and cyber-attack	• Cloud security • Data-centric security
Application layer	• Privacy • Application designed in secured manner • Policies and standards acceptable worldwide	• Data corruption • Cross-site scripting (XSS)	• Connected system • Service management
Business layer	• Protect privacy and allow the flow of information in a hierarchical model as proposed by the company's policies • Responsible for maintenance and adaptability	• Developer's cookie tampering and business process/logic bypass • Business constraint exploitation	• No leaking of data outside the organization

3.4.1 Recognition Layer

The recognition layer, also called the things layer, is the first and foundational layer of 5G-oriented IoT architecture. This layer is a combination of hardware and physical objects. The various attacks to which this layer is subject are:

False Information. Transmitting incorrect information on the wireless network is an example of false information.

Timing. The sender's information must be synchronized with the receiver's information within the specified time duration.

Jamming. In a jamming attack, the intruder sends a high range of radio signals to disrupt communication in the network to minimize the SINR (signal-to-interference-plus-noise ratio).

Tampering. This causes physical damage or destruction to parts of the device such as the sensor node.

Several hardware and software security actions are required to protect the data from attacks in the recognition layer. First, sensor data must be protected from unattended access. Second, a key agreement is required for data transmission to authorized personnel. Third, physical devices should be protected from malicious attacks such as jamming and tampering.

Hardware Security. Various parts of the hardware such as sensors and microcontrollers are vulnerable to attacks such as jamming and tampering. Intelligent security actions in the recognition layer provide hardware security.

Software Security. Software is also vulnerable to various attacks that can disrupt the integrity of the network by altering the data. The threat of false information is prevented by a key agreement concept between the two parties. The combination of public and private key agreements provides software security in the recognition layer.

3.4.2 CONNECTIVITY LAYER

This is the second layer of the 5G-oriented IoT architecture, also known as the edge computing layer. Its main function is to provide connectivity and edge computing via communication protocols and networks. Attacks in the connectivity layer degrade the performance of the network and include:

Man-in-the-middle attack (MITM). MITM is an attack in which an intruder joins the network as a relay between the communicating parties and alters the message transmitted.

Spoofing. A spoofing attack is defined as an attack in which an intruder passes false information in the network on behalf of another node and creates chaos in the network.

Eavesdropping. This is an attack in which an intruder inserts a relay in the network and gathers insecure information being communicated between two parties.

Sinkhole. A sinkhole is an attack in which a malicious node tries to attract traffic or neighboring nodes by sending false information on new routes and new routing tables present in the network.

Sybil. This type of attack in the network uses multiple false identities to gain influence and subvert the reputation system.

The network security actions required in the second layer are: (i) to provide integrity to a document; (ii) to provide encryption techniques within the solution for existing ID mechanisms; and (iii) optimized communication protocols for high-speed data transmission in the 5G-oriented IoT scenario.

3.4.3 SUPPORT LAYER

The support layer is the third layer of 5G-oriented IoT architecture; it provides a support platform to both downward and upward layers. Cloud computing, intelligent computing, and data analysis functions are performed in the support layer. The main attacks affecting the support layer are:

Malicious Insiders. A malicious insider is a person who misuses legitimate access of an organization and disrupts the confidentiality of the information.

Viruses. A virus is defined as a malicious code that replicates itself in the computer without the permission of the authorized user.

Account Hijacking. This is an attack in which a cloud account or data-centric account are hijacked by the intruder.

Cloud Malware Injection Attack. This is an attack in which an intruder tries to add a malicious service to the cloud.

Hacking and Cyber-attack. This is an attack in which an intruder takes control of the system for communication and maliciously disables it to steal the information.

Important security actions in the support layer are antivirus protection, cloud security and encryption.

Attacks that occur in the cloud include account hijacking and cloud malware injection attacks. These issues are solved using encryption techniques with the existing solutions. Intelligent secured actions provide cloud security in the support layer. Hijacking attacks and cyber-attacks can also occur in the data center, and similarly, intelligent secured actions are provided for the data center. Cryptography helps to provide data-centric security.

3.4.4 APPLICATION LAYER

The application layer is the fourth layer of the 5G-oriented IoT architecture. This is the terminal layer that provides diverse applications and services. Attacks present in the application layer are:

Data corruption. In this type of attack a malicious node corrupts the data. When the sender transmits the data, an intruder corrupts the data before it reaches its destination.

Cross-site scripting. This type of attack occurs in web applications. The intruder injects a malicious script at client level to bypass access controls.

Security actions providing security in the application layer in terms of connected systems and service management are: (i) provide privacy to the users; (ii) secure application design; and (iii) policies and standards that are acceptable worldwide. Attacks injected into the network, such as cross-site scripting, can be prevented using ID mechanisms introduced at the design stage of the application layer. Network service management can be disrupted due to data corruption. These issues can be overcome by hiring educated professionals to maintain and manage the application of the 5G-oriented IoT.

3.4.5 BUSINESS LAYER

The business layer, the last layer in the architecture of 5G-oriented IoT, creates business models and manages all IoT systems. Attacks occurring in this layer are:

Tampering with developers' cookies and business process/logic bypass. Website developers use session cookies, build data using session-only variables, and set cookies in the browser to expose logic holes. These cookies can be re-engineered and their value predicted by intruders exploiting the logic by altering the value of cookies.

Business constraint exploitation. Business logic should have defined rules and constraints. If it is not defined properly or poorly designed, intruders can easily attack and alter hidden logic fields. Intelligent secured actions in the business layer to prevent data leaking outside the organization are: protection of privacy to allow the flow of information in a hierarchical model as proposed by the company's policies; and maintenance and adaptability.

3.5 INTELLIGENCE IN THE 5G-ORIENTED INTERNET OF THINGS

The interconnection of IoT and 5G technology paves the way for the production of huge volumes of data from sensors and systems. AI and more specifically machine learning (ML) allows users to optimize the use of network resources efficiently. The ML techniques are of three types: supervised, unsupervised and reinforcement learning [18]. In supervised learning, a dataset containing labeled data is available for the training of the model. Unseen data can, later on, be classified based on the learning gained by the ML model according to the existing dataset. Unsupervised learning uses unlabeled data and works on the principle of finding the hidden patterns in the data. This type of learning algorithm is mainly used for description-based learning paradigms. Reinforcement learning is a learning process based on rewards and penalties. It is also known as semi-supervised learning.

In a real-time IoT–5G integration scenario [19], a lot of data is generated and used to make systems more adaptive. This also helps to optimize the system based on a single decision factor or multi-objective optimization issues.

3.6 TOOLS

In view of the vast amount of data accumulated by the bridging of the two popular technologies, IoT and 5G, specialized statistical methods and are important [20, 21]. Some of these tools and programming languages are discussed in this section.

3.6.1 HADOOP

Hadoop (Figure 3.4l) is an open-source software framework used to execute applications on numerous interconnected computational devices and to store enormous amounts of data. It contains built-in libraries for big data analysis and modeling. The main benefit of Hadoop is that it can cluster multiple computers in parallel and with improved speed. Hadoop is scalable if there is a considerable increase in data volume. Since it is a Java-based platform, it is compatible with all the major existing operating systems.

3.6.2 SPARK

Apache Spark (Figure 3.5) is similar to the Hadoop framework and uses a cluster-based environment and in-memory calculations. The processing response time is better and it is quite easy to learn. It can easily be integrated with Java, R and Python programming languages.

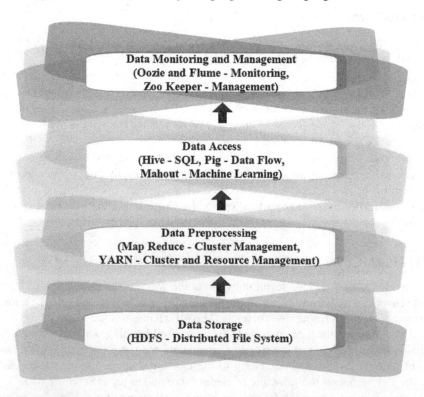

FIGURE 3.4   The Hadoop framework.

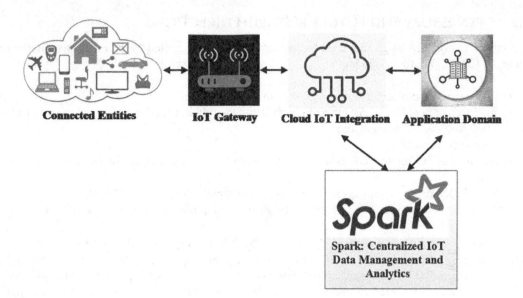

FIGURE 3.5 Apache Spark.

3.6.3 Hive

A hive is a popular tool for data integration. It has an extension named HiveQL which is used to run queries similar to SQL on the data stored in databases. The Hive Metastore contains all the metadata linked with the tabular structure. It is best for online transactions and stores data as plain text files. The main task carried out by Hive is data analysis.

3.6.4 R Studio

Demand for coding languages is increasing daily. R is an open-source coding language used to perform statistical analysis on datasets. The main advantages of R are fast processing of large volumes of data and the successful application of ML techniques for data exploration and prediction. Exploratory IoT evaluations can be easily carried out using R.

3.6.5 Python

Due to its flexibility and dynamic nature, developers prefer Python for designing intelligent IoT devices. "Python is the language of choice for one of the most popular microcontrollers in the market, the Raspberry Pi," according to Kinmay Covey, a popular microcontroller developer [24]. Business logic is directly implemented on the device using this programming language.

3.6.6 CupCarbon

CupCarbon is an IoT and wireless sensor network (WSN) simulation tool. It can help in the creation of two-dimensional or three-dimensional designs, and can visualize data flow patterns, radio signal propagation and interference in IoT. Its main purpose is to develop smart cities, hence its utmost importance to the entire world. This simulation tool is very efficient as it is based on core performance parameters such as security, power, and energy.

3.7 OPEN ISSUES AND FUTURE RESEARCH DIRECTIONS

Among the many open issues that need further exploration as suggested by the exhaustive literature surveyed [22, 23] are the following:

- The impact of molecular absorption on path loss and noise.
- Research in novel transmission techniques using the relationship between transmission bandwidth and transmission distance.
- Terahertz channel modeling.
- Accurate and efficient prediction of energy consumption and power parameter in 5G-integrated IoT.
- Security issues at various layers of the integrated architecture.
- Novel intrusion detection mechanisms for distributed denial of service attack (DDoS attack).

In summary, the main areas covered by this integrated technology are infotainment (smart wearables, robotics), smart home (intrusion detection, energy management, access control), smart city (traffic control, waste management), retail (fraud detection, traffic tracking, smart payments), agriculture (soil monitoring, greenhouse monitoring), and environment monitoring (air and water quality, drone surveillance). Future research directions can be established according to the challenges presented in each domain.

3.8 CONCLUSION

This chapter has outlined a novel layered integration framework with intelligence and security features that enable more efficient information processing and optimization. Functional and non-functional quality of service design properties of this framework are discussed with special emphasis on the security parameters for each layer. Various tools used for IoT, 5G, and data analytics for intelligent computing are discussed. The most popular languages used for data analytics nowadays are R programming and Python. The principal security and intelligence properties are reported, with possible future research directions in the domain of smart, secure and intelligent computing based on extensive data analytics.

REFERENCES

1. Varga, P., Peto, J., Franko, A., Balla, D., Haja, D., Janky, F., & Toka, L. (2020). 5g support for industrial IOT applications–challenges, solutions, and research gaps. *Sensors*, *20*(3), 828.
2. Shafique, K., Khawaja, B. A., Sabir, F., Qazi, S., & Mustaqim, M. (2020). Internet of things (IoT) for next-generation smart systems: A review of current challenges, future trends and prospects for emerging 5G-IoT scenarios. *IEEE Access*, *8*, 23022–23040.
3. Wang, D., Chen, D., Song, B., Guizani, N., Yu, X., & Du, X. (2018). From IoT to 5G I-IoT: The next generation IoT-based intelligent algorithms and 5G technologies. *IEEE Communications Magazine*, *56*(10), 114–120.
4. Li, S., Da Xu, L., & Zhao, S. (2018). 5G Internet of Things: A survey. *Journal of Industrial Information Integration*, *10*, 1–9.
5. Wong, V. W. (Ed.). (2017). *Key technologies for 5G wireless systems*. Cambridge University Press.
6. De Almeida, I. B. F., Mendes, L. L., Rodrigues, J. J., & da Cruz, M. A. (2019). 5G waveforms for IoT applications. *IEEE Communications Surveys & Tutorials*, *21*(3), 2554–2567.
7. Zhong, M., Yang, Y., Yao, H., Fu, X., Dobre, O. A., & Postolache, O. (2019). 5G and IoT: Towards a new era of communications and measurements. *IEEE Instrumentation & Measurement Magazine*, *22*(6), 18–26.
8. Rusti, B., Stefanescu, H., Iordache, M., Ghenta, J., Brezeanu, C., & Patachia, C. (2019, June). Deploying Smart City components for 5G network slicing. In *2019 European Conference on Networks and Communications (EuCNC)* (pp. 149–154). IEEE.

9. Rahimi, H., Zibaeenejad, A., & Safavi, A. A. (2018, November). A novel IoT architecture based on 5G-IoT and next generation technologies. In *2018 IEEE 9th Annual Information Technology, Electronics and Mobile Communication Conference (IEMCON)* (pp. 81–88). IEEE.

10. Ahmad, I., Kumar, T., Liyanage, M., Okwuibe, J., Ylianttila, M., & Gurtov, A. (2018). Overview of 5G security challenges and solutions. *IEEE Communications Standards Magazine*, *2*(1), 36–43.

11. Ye, Q., Li, J., Qu, K., Zhuang, W., Shen, X. S., & Li, X. (2018). End-to-end quality of service in 5G networks: Examining the effectiveness of a network slicing framework. *IEEE Vehicular Technology Magazine*, *13*(2), 65–74.

12. Chettri, L., & Bera, R. (2019). A comprehensive survey on Internet of Things (IoT) toward 5G wireless systems. *IEEE Internet of Things Journal*, *7*(1), 16–32.

13. Sodhro, A. H., Obaidat, M. S., Abbasi, Q. H., Pace, P., Pirbhulal, S., Fortino, G., & Qaraqe, M. (2019). Quality of service optimization in an iot-driven intelligent transportation system. *IEEE Wireless Communications*, *26*(6), 10–17.

14. White, G., Nallur, V., & Clarke, S. (2017). Quality of service approaches in IoT: A systematic mapping. *Journal of Systems and Software*, *132*, 186–203.

15. Wang, N., Wang, P., Alipour-Fanid, A., Jiao, L., & Zeng, K. (2019). Physical-layer security of 5G wireless networks for IoT: Challenges and opportunities. *IEEE Internet of Things Journal*, *6*(5), 8169–8181.

16. Fang, D., Qian, Y., & Hu, R. Q. (2017). Security for 5G mobile wireless networks. *IEEE Access*, *6*, 4850–4874.

17. Mavromoustakis, C. X., Mastorakis, G., & Batalla, J. M. (Eds). (2016). *Internet of Things (IoT) in 5G mobile technologies* (Vol. 8). Springer.

18. Chettri, L., & Bera, R. (2019). A comprehensive survey on Internet of Things (IoT) toward 5G wireless systems. *IEEE Internet of Things Journal*, *7*(1), 16–32.

19. Wang, D., Chen, D., Song, B., Guizani, N., Yu, X., & Du, X. (2018). From IoT to 5G I-IoT: The next generation IoT-based intelligent algorithms and 5G technologies. *IEEE Communications Magazine*, *56*(10), 114–120.

20. Mavromoustakis, C. X., Mastorakis, G., & Batalla, J. M. (Eds). (2016). *Internet of Things (IoT) in 5G mobile technologies* (Vol. 8). Springer.

21. Najm, I. A., Hamoud, A. K., Lloret, J., & Bosch, I. (2019). Machine learning prediction approach to enhance congestion control in 5G IoT environment. *Electronics*, *8*(6), 607.

22. Mumtaz, S., Bo, A., Al-Dulaimi, A. and Tsang, K.F., 2018. Guest editorial 5G and beyond mobile technologies and applications for industrial IoT (IIoT). *IEEE Transactions on Industrial Informatics*, *14*(6), 2588–2591.

23. Chowdhury, M. Z., Shahjalal, M., Hasan, M., & Jang, Y. M. (2019). The role of optical wireless communication technologies in 5G/6G and IoT solutions: Prospects, directions, and challenges. *Applied Sciences*, *9*(20), 4367.

24. Zaidi, S. M. A., Manalastas, M., Farooq, H., & Imran, A. (2020). SyntheticNET: A 3GPP compliant simulator for AI enabled 5G and beyond. *IEEE Access*, *8*, 82938–82950.

4 Advances in Mobile Communications from a 5G Perspective

Anuradha Singh, Ayushi Tandon, and Laasya Cherukuri
Indian Institute of Information Technology, Nagpur, India

CONTENTS

4.1   INTRODUCTION

With the growing ambition of making life simpler, the application of electronic gadgets is now rising exponentially. The vast number of users can no longer be managed by the limited spectrum allowance of 4G networks. The services provided by smartphones are significantly increasing mobile traffic,

DOI: 10.1201/9781003045809-5

and support is needed for low-cost rapid and reliable high-bandwidth services. Furthermore, the provision of internet service to rural regions and the convergence of disparate networks on a single interface remain challenging.

All these factors have led to the establishment of the next-generation wireless communication known as 5G. Fast data rates with low bandwidth and limited energy consumption can be successfully achieved using this technology, which is also capable of managing product responsibility. Despite its many advantages, there are still many implementation and performance issues in 5G technology. Significant research and development to support high mobile data traffic on 5G communication networks is ongoing. While spectrum limitations may also be solved by dynamic spectrum distribution, researchers are investigating the feasibility of using the idle spectrum in a range of 30~300 GHz millimeter waves to increase bandwidth [1]. Among the many advances proposed and achieved in the existing fundamental units of the network are cognitive radios and small cells. IoT features such as cloud computing and artificial intelligence distinguish 5G from its predecessor 4G as stepping stones in the overall development of mobile communications.

This chapter explores the functionality of the cognitive radio system as one of the primary drivers for the forthcoming 5G mobile network. The deployment of additional bandwidth is a constructive way to increase the data rate and system efficiency of 5G networks. The chapter presents the basics of CR, spectrum optimization in 5G technology, carrier aggregation with reference to cognition in 5G networks, the primary structure required and barriers to the deployment of 5G technology.

As we move into the 5G age, enhancements in small-cell technologies along with compatible approaches such as massive MIMO, mm-wave propagation and supplementary bandwidth theoretically enable the 5G mobile approach. Though small-cell technology is still undeveloped, if we can overcome challenges like interference coordination, this technology can prove a strong starting point for achieving the 1000x challenge [2].

Other small-cell advances include the software-defined radio (SDR) system which addresses expected performance standards by providing an improved approach to adaptability and functionality, and multicore DSP, which provides increased computing capacity to develop multi-functional small cells with defined embedded processors to significantly boost performance [3].

Massive growth has been seen in the development of mobile communications by the fundamental IoT architecture incarnation. Technologies like semantics, cloud services, sensor hardware/firmware, data modeling, and so on, offered by IoT have provided a new approach to the development of 5G networks.

It is expected that 5G will expand beyond conventional voice and data networks. The increased reliability, lower latency and higher network characteristics connecting throughput and increased density will enable massive commercial activity. Deployment of innovations such as IoT, AI, AR/VR, will enhance the use cases around markets such as manufacturing, communications, culture, education, sales, production, and agriculture.

5G networks are expected to give rise to innovative product markets with many new mediators focused on accessibility and service components and software in the value chain of telecom internet providers. Collaboration across the industry may be critical to enable this technological innovation all across the production chain, but it will surely be implemented in the best way possible [2].

4.2 COGNITIVE RADIO PERSPECTIVES

5G and cognitive radio (CR) are the two technological advances designed to fulfill the high growth of network data traffic in wireless communication systems. In the future, the latest age of connectivity will be governed by 5G. Because the future of mobile broadband will primarily be powered by superfast-definition content, and as everything around us will still be connected, 5G aims to provide 10 GB/s of increased bandwidth and network speed.

The primary purpose of cognitive radio is to coordinate a more productive application of the spectrum, as it adapts to present an optimal medium of communication [4]. Thus, advances in cognitive radio itself will play a vital role in the advancement of 5G in the future.

Implementing CR will improve the present overcrowding of the continuum. Currently, 15–85 percent of the licensed spectrum is so often unused that it leads to inefficient usage of the current network spectrum, with many licensed users not completely using the spectrum assigned to them [4]. The license provides users with access to transmit and receive wireless signals, and cognitive radio makes use of the underutilized spectrum by making it accessible to the secondary system (i.e., non-licensed users) without adversely affecting primary users. CR recognizes the spectrum's white blanks based on different factors like time, frequency or geographic location [5].

4.2.1   CR FUNCTIONALITIES

The performance of a spectrum is the building block of overall wireless communication, and heavy data traffic now needs to be accommodated. 5G broadband networks have to have high data traffic and are marked by intensive reuse of bandwidth, intense distribution of ground stations and wireless devices, and incorporation of several networking technologies. In CR, the efficiency of the server is managed using a variety of channel management and synchronization methodologies.

The architecture of CR is such that all its layers are segregated to build a rush-free network with different layers each having its own unique functionality [6]. The physical layer uses various spectrum detection methodologies. The link layer accounts for the classification and monitoring of the radio system. Bandwidth-conscious configuration is handled by a network layer, and bandwidth is handled by a transport layer. Both the specifications of the user functionality and QoS are managed by the application layer.

Key functions of cognitive radio have been used to enable dynamic spectrum access. They are:

- Spectrum sensing
- Spectrum management and decision
- Spectrum sharing
- Spectrum mobility.

Taken together, these functionalities provide wide scope for the enhancement of the spectrum and hence development in wireless networks.

4.2.1.1   Spectrum Sensing

This contributes to the precise identification of spectrum spaces. It must be consistent in such a way that once the PU re-accesses the spectrum, CR nodes are automatically instructed to cease communication [6]. It can be applied through in-band detection, out-of-band detection and geolocation databases. It is also intended to modify external characteristics, such as power ratings, passwords, and so on.

4.2.1.2   Spectrum Management

As multiple spectrum holes are spread across a large frequency band, spectrum management is supposed to choose the most appropriate one available for spectrum sharing. The decision is made by evaluating transmission power, bandwidth, modulation techniques, coding schemes and scheduling. Selection also varies depending on the quality of service (QoS) criteria for CR development requirements, such as packet error rate, latency, etc. [1].

The spectrum decision is made after the spectrum blank spaces or holes are categorized based on parameters like potential interference on the primary network, mutual CR interference, channel capacity, frequency band and holding time.

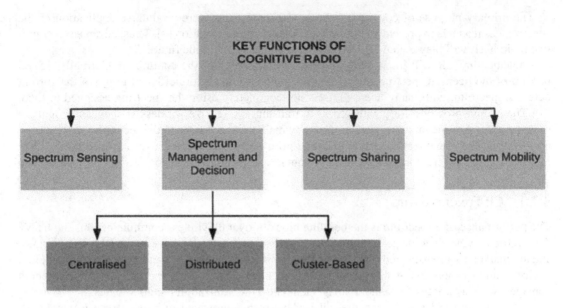

FIGURE 4.1 Spectrum key functions of cognitive radio.

After each of these parameters has been taken into account, there are three different ways in which spectrum access decisions can be made (Figure 4.1):

- Centralized
- Distributed
- Cluster-based

4.2.1.3 Spectrum Sharing

Spectrum sharing means an equal distribution of spectrum holes for various CR applications. It is centered on planning involving time, frequency, code and spatial measurements. It also prevents unintended intra-network involvement. Spectrum sharing can be categorized into two methods: centralized and distributed [1].

In centralized spectrum sharing, a main body monitors accessibility and selection using the effects it detects and collects from the decentralized nodes. The body determines the frequency judgments. In distributed sharing, each node takes spectrum decisions depending on its own knowledge and laws without the involvement of the central body [5].

4.2.1.4 Spectrum Mobility

This indicates the capacity of nodes in the network to move between various spectrum holes smoothly, according to the circumstances. This would include PUs re-accessing the unused spectrum, unacceptable channel conditions within the existing range of frequencies, or raising bandwidth to meet the increasing need for connection speed. The switchover between the various spectrum holes is called the spectrum handoff.

4.2.2 COGNITIVE RADIO IN 5G

Looking at the features and attributes of 5G and CR, there seem to be several parallels between the two technologies. The key connection is that they are able to interact among several devices or systems, and are also receptive to the idea of modifications and handling new versatile policies.

Security issues remain a major concern with both 5G and cognitive radio in the current evolving phase of the innovation [1]. In a nutshell, 5G aims to incorporate different cellular technologies. Cognitive radio is a versatile device enabling auto-assimilation into a wide spectrum of wireless domains, making it very versatile and appropriate for use as a 5G terminal [5]. A 5G terminal has such competence that it could actually operate with various types of cellular networks and switch to multiple theories over the same sessions.

The following sections describe advances that can be further modified in the field of cognitive radio technology to boost it up to the maximum for 5G usage [7].

4.2.2.1 Antennas for CR in 5G

Modern technologies can be integrated with compact antennas and MIMO applications.to provide ultra-wide spectrum and high-performance efficient antennas. Recent advances in cognitive radio have illustrated the requirement for innovative and upgradable antenna systems with improved performance. Another advantage of antennas controlled for 5G is their small size, since 5G emphasizes the application of the millimeter-wave (mm-wave) spectrum. Due to this, there is a significant decrease in the antenna gain as it is directly proportional to the aperture. Several antenna array architecture configurations considered appropriate for 5G technology have been integrated, in which circular arrays are used to have a flat gain [2]. This is still not reasonable for 5G bands since it reaches up to 40 GHz. A wide-band magneto-electric antenna with good bandwidth frequency response from 26.5GHz to 38.3GHz was suggested in [8]. Also, the antenna is equipped with a fork-shaped ridge gap waveguide for low power loss and improved or increased bandwidth at mm-wave frequencies [9]. Thus, the incorporation of the MIMO technologies in CR antennas has brought advancement in mobile communications up to the next level.

4.2.2.2 Cognitive Engines

Knowledge of CR is derived from the existence of cognitive engines. A cognitive engine can be called the brain of CR, with skill set, learning, and logic-engine components. The engine has the capacity both to learn and reason, and to store gained skills dependent on the learning methods in the skill set (see Figure 4.2). Thus, several advances have been achieved in CR engines to meet the requirements of the proposed 5G network.

Cognitive engine architecture, focused on the cognition cycle, uses specific techniques such as genetic algorithms, adaptive mutation mechanisms, different types of bio-inspired algorithms and computational learning. FPGA (field programmable gate arrays), MIMO, FFT-based processing units and nanoscale processing are among the innovative methods that may be used in the advancement of the cognitive engine.

4.2.2.3 Improved PHY Technologies

The physical layer of cognitive radio is described as having multiple carrier technologies to accelerate the performance of the system [6]. These are:

OFDM-QAM (Orthogonal Frequency-Division Multiplexing Quadrature Amplitude Modulation). This is a type of modulation with perpendicular subcarriers, and every subcarrier is amplified with a quadrature amplitude modulation scheme [6].

OFDMA (Orthogonal Frequency-Division Multiple Access).This refers to a multiple-user network access technique in which each individual is allocated a list of subcarriers and also provides varied QoS levels [6].

MC DS CDMA (Multi-carrier Direct-Sequence Code Division Multiple Access. This refers to a multi-user system in which each operator has a particular protocol and can use several parallel subcarriers.

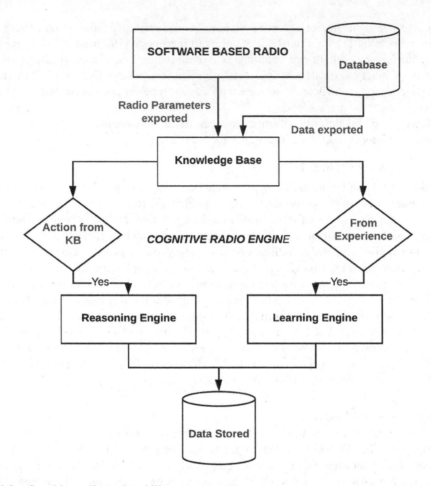

FIGURE 4.2 Cognitive radio engine skill-set components.

5G is implementing (although with minor variations) the suggested CR standards as well as improved performance strategies to present a sustainable and efficient wireless networking system.

4.3 SMALL-CELL COVERAGE IN 5G COMMUNICATIONS

With the expansion in mobile broadband communications followed by demand from subscribers, a massive 1000x increment in mobile data traffic is expected in the next decade. To satisfy the antici-pated traffic, 5G innovation should be empowered to minimize the deployment cost and to maintain fast and financially intelligent data connectivity. Under the present innovation guidelines, there has been greater emphasis on the need for more spectrum (increasing spectral efficiency) and high con-centration of cells per unit area [10].

With evolving technologies, the possibility of imparting confined resources while maintaining a satisfactory service quality through the stationing of small cells has proved an attractive solution.

Strictly defined, small cells are cellular nodes consuming less power which work within an autho-rized range and are supervised by the provider to support an improvised cellular network.

4.3.1 TRENDS IN SMALL CELLS

As indicated by market reports and some industrial experts, small cells in residential areas are satu-rated but there is growth in enterprises and public areas. Small cells are also an integral part of heterogeneous networks (HetNets) seeking to increase capacity and bandwidth utilization while focusing upon lowering the cost per bit of data transport [3].

Although HetNet architecture is very primitive, its scope goes beyond small cells to encompass a multitude of architecture, layers and radio access technology (RAT) styles which require coexistence and mutual assistance. Moreover, boosts to integrated and advanced technologies are needed to handle interference, various modes of traffic and advanced services. Several technologies have been introduced in small 5G networks, for example carrier aggregation, authorized and unauthorized band, programming and amplification, multi-cell cooperation and advanced MIMO [3].

4.3.2 TECHNICAL ASPECTS OF SMALL CELLS

4.3.2.1 Carrier Aggregation

In order to satisfy multi-GB/s data limit criteria, small cells need to use larger bandwidth, for which carrier aggregation is a key technology. Carrier aggregation aims at achieving data rates by combining various LTE component carriers (CCs) so as to provide wider bandwidth for both downlink and uplink. The component carrier can have a frequency band of 1.4 to 20 MHz. In total only five CCs can be integrated, thus the maximum bandwidth that can be obtained is 100MHz. [11].

Thus, greater flexibility can be provided to the aggregation for covering different variations. Multi-RAT aggregation such as LTE and WiFi integration can also be supported. In addition, the dual connection of access points to numerous small cells will result in the deployment of aggregated spectral region in various small cells (Figure 4.3).

4.3.2.2 Multi-Cell Cooperation

The main problems for the advanced distribution of small cells in the upcoming 5G technology are congestion regulation and space allocation. Recognizing these limitations, a wider range of IoT disturbance control strategies has been made accessible to 4G-dependent small-cell networks, including backhaul-based coordination, sub-band scheduling, and adaptive fractional frequency reuse [3].

4.3.2.3 Massive MIMO

The MIMO technique improves spectrum efficiency to provide an alternative to enhancing system capacity. The initial network concretion strategy includes extensive installation of antennas per cell site, to form a 'massive MIMO' i.e., MIMO system with huge antennas.

As the opening of the array expands to accommodate more antennas, its resolution often improves. This essentially concentrates the transmission power to the intended recipients, so that transmitting may be arbitrarily low, resulting in substantial decreases in (and eventually complete elimination of) intra- and inter-cell interference. The concentrated installation of small cells is an alternative approach to multi-antenna technologies. This is reflected in clients particularly close to the antennas obtaining substantial improvements in bandwidth and has been recognized as a successful way to maximize network efficiency, since power corresponds linearly to the density of the cells [10].

Together with channel estimation and interference mitigation techniques, this methodology leads to improved efficiency of the spectrum.

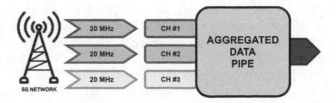

FIGURE 4.3   Carrier aggregation in 5G.

4.3.2.4 Multiple Access Techniques

In order to satisfy the increasing needs of emerging communication technologies, a remarkable move to better frequency ranges like millimeter-wave frequencies (mm-wave) is occurring.

OFDMA (orthogonal frequency-division multiplexing access), which is OFDM's multiple-user design, is commonly utilized in current 4G technology. Although it has key advantages, the main limitation making it quite inappropriate for 5G networks is that every subcarrier in an OFDM network includes a rectangular-shaped panel in the time domain, resulting in a sinc-shaped subcarrier in the frequency domain [12].

The space division multiple access (SDMA) concept has been developed separately from the widely used time and frequency multiple access network. Compared to conventional mobile communication systems, radio bandwidth can be improved in an SDMA-based network by utilizing antenna arrays on the side of the base station (BS) and in due course directing the signals in uplink as well as downlink.

These methods offer the ability to support several users at a time by overlaying the beams in a specific spectrum, enabling the system's capacity to be enhanced. The importance of SDMA is not merely its multi-user capabilities; it also improves the performance of the spectrum by allowing frequency re-utilization inside a small cell [12].

4.4 SMALL CELLS AND 5G

The saturated deployment of small cells is the next innovation for network infrastructure. Improvement is required to promote increased internet speeds and reduce the lag on multi-cell synchronization, fronthaul and backhaul.

The radio access network (RAN) framework for the upcoming 5G technology is designed to support a somewhat more standardized level of operation around the whole coverage range, reducing traditional performance problems based on user position.

However, it is possible to deploy ground stations with a centralized RAN (C-RAN) network that includes a centralized unit (CU) as well as multiple local distributed units (DUs) to increase operating performance. In the base-band units (BBUs) of these various DUs, each DU in a C-RAN system is fitted with radio front ends to include antennas, RF circuitries and fully integrated components [13].

The system divides the base station into two entities which are adjustable at multiple places and controlled by the configuration of the fronthaul. In addition, since the BBUs of several radio stations are situated in the central unit, advanced coordination-based network systems are now much more feasible, as strong collaboration between several radio sites is now enabled [13].

Due to the predicted existence of both non-distributed and distributed centralized ports in the RAN structure (Figure 4.4), the transport network becomes more advanced, since both backhaul (interconnection between core network and access nodes) and fronthaul (interconnection between CUs and DUs of C-RAN) need interconnections for multiple forms of network connections/nodes [13].

4.5 IoT PERSPECTIVE

This section deals with the application of cognitive radio networks and small cells in the IoT. It was Kevin Ashton who coined the term 'Internet of Things'. The concept of IoT is to connect all devices to the internet enabling them to communicate with each other all over the world. IoT can be considered as a massive network of devices, vehicles, buildings and others that are equipped with sensor circuits and network connections that allow things to exchange their data. An increase in the rate of change in technology has led to the enhancement of IoT through collaboration with the latest and upcoming technologies. Here we consider the fundamental and theoretical concepts of the 5G network with respect to IoT.

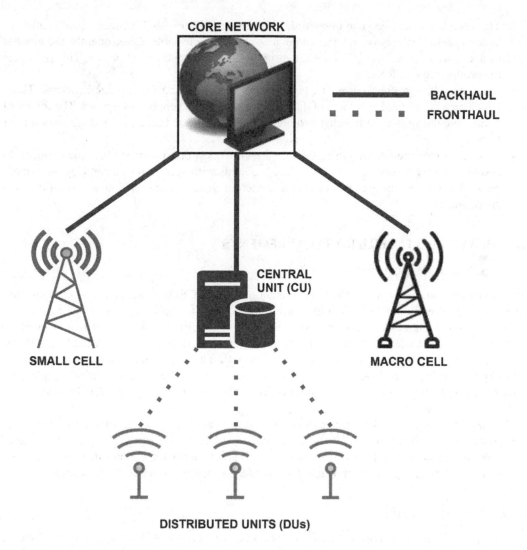

FIGURE 4.4   Backhaul and fronthaul in RAN.

The functionalities of cognitive radio networks (CRNs) are spectrum sensing, spectrum deci-sion, spectrum management and spectrum mobility. No comprehensive report exists in the literature describing how all these features are going to help in enhancing IoT.

The functionalities of small cells are carrier aggregation, multi-cell cooperation and massive MIMO.

4.6   DIRECTING CRNS TOWARD IoT

The building of the IoT is involving numerous technologies, including cloud computing, sensor deployment, data modeling, storage analysis, and so on. IoT networking systems are gradually mov-ing from wired to cellular: wired such as digital subscriber line (DSL), ethernet; or wireless such as WiFi, satellite, and so on. All these are of short range and have some limitations [14].

The following factors explain the need to incorporate cognitive radio infrastructures into IoT devices:

1. Wireless technologies are obsessed with communication between sensors and cloud services. Wireless systems including Bluetooth and Zigbee have such a restricted range of frequencies.

2. The situation would become impossible for a large number of IoT artifacts, as allocation of bandwidth to this massive number of devices will not be feasible. Consequently, the number of authorized clients would increase, causing issues for unauthorized access. This promotes the production of CRNs.
3. Users only like to use devices that have no real network access or storage concerns. These CRNs provide a massive amount of data so massive storage support is required. The theory of cloud computing is already being explored as computers send data to cloud servers on their own.
4. Typical wireless networking infrastructure does not allow consumers to share the network. In addition, future mobile networks will face problems with spectrum sharing. Cognitive radio networks may be helpful as there could be numerous devices, each seeking accessibility to the spectrum.

4.7 HOW CRNS FULFILL IoT REQUIREMENTS

4.7.1 CHANNEL ALLOCATION

CRNs must have at least one interface for interaction in IoT. Channel allocation regulations have to some extent already been explored in cognitive radio, yet its application to IoT devices is fascinating. Research suggests an opportunistic strategy for the distribution of cognitive channels in wireless IoT. For channel estimation, a traffic background support algorithm is used, distributed very efficiently with minimal interference with heavy traffic history interconnection. In general, methodology has intervened rapidly in transitions in geometry and connections which might arise in the system [14]. It provides appropriate channels for the dissemination of content for unexpected cognitive radio networks.

This approach considers the occupancy of PU channels and analysis of social networks to pick suitable channels and neighbors for dissemination of content. The study analyzes PU channel utilization and system design to choose the appropriate channels for content dissemination. We assume that incorporating this method into CR-based smart nodes would be the best solution.

4.7.2 PROTOCOL DESIGN

Development of cognitive radio-based IoT protocols is also increasingly essential. Protocols should not only fulfil the demands of QoS, but should also secure network nodes. At the same time, these parameters from PHY to the transport layer must be capital intensive, secure and web enabled. Scientists and researchers see cognitive M2M configurations focused on radio as the framework for future IoT.

4.7.3 ENERGY HARVESTING

Cognitive radio nodes could also function when upholding their level of operation and must therefore be applied to energy harvesting schemes. It is also possible that a remote-control wireless hosting node would be considered for information retrieval, or nodes could even be fitted with a cellular voltage supply.

4.8 SMALL CELLS FULFILLING THE IoT REQUIREMENT

Traffic optimization and execution are two main benefits when we combine small cells and IoT. Small cells provide area coverage, which ensures that they provide electricity where a need occurs. As they seem to be nearer to an individual, they demand hardly any radio frequency power to supply large bandwidth. In small cells, the type of information communicated has various limitations and is prioritized accordingly.

The way that security protocols are implemented will therefore adhere to the protection and privacy required for that specific packet. The various rates of connectivity need to be identified before the small cell can broadcast them. Providing end-to-end security and authentication services for the network is perhaps the most important aspect of any small-cell operation.

Different services in IoT systems can use small cells both to minimize data traffic from the traditional cellular layer and even to increase the coverage and efficiency of the entire network. The use of small cells in IoT will significantly increase the energy efficiency of the system. With the implementation of small cells, more users on the same radio spectrum can be packed into a given area, allowing for higher spectral efficiency of the area.

4.9 SMALL-CELL DEPLOYMENT THROUGH CR

CR continues to focus on dynamic spectrum access, which had already succeeded in providing an effective solution to connectivity. CR does have a beneficial impact on network interference control, the level of energy usage and lifespan, and can be implemented in some small IoT-based cells. The femtocell base station (FBS) is a small base station typically designed for indoor use in a small office home office (SOHO) environment to provide voice and broadband services. FBS is a cost-competitive, low-power, short-range wireless device operating in the licensed spectrum. FBS also utilizes capacities such as bandwidth detection and empty band identification for using low-end UE spectrum deployment and handling network limits. CR's importance lies in the assumption that there is a correlation between power consumption and frequency bandwidth, which implies that if we want to own more bandwidth, we have to reduce power consumption [15].

To put it another way, we want to have dynamic and optimal spectrum management—which is where CR has a vital role to play. A study of CR-supporting structures and techniques would be helpful in proposing more cost-effective and/or complicated strategies [16, 17]. CR networks may also be used as an effective way to solve bandwidth depletion by improving spectral quality with sufficient use of bandwidth [17]. The productivity of CR networks therefore very much depends on primary radio user activity [17]. Appropriate study and analysis is therefore essential in order to effectively organize the activity of primary users in CRNs.

Furthermore, several network/data usage surveys are often carried out in a specified location to obtain accurate spectrum usage data [18]. Research into alternatives and strategies for spectrum occupancy compilation and control would therefore help to get a more real dataset of spectrum usage in order to produce a framework to manage spectrum-related improvements. Small cells with CR functionality can also assign radio resources in an improved way. They will also help reduce intervention from co-channel and advance spatial reuse [19].

4.10 CONCLUSIONS

This chapter has provided an overall summary of 5G technology and the incorporation of different technologies into the 5G network to address the future demands of 5G. This system offers a wide variety of capabilities while enabling higher network data traffic. Advances in technology have been made with the help of heterogeneous networks of small, micro and macro cells. Cognitive radio is regarded as the backbone of mobile communications with improved technologies. In the search for an ultimate 5G network, IoT-based 5G is a thriving aggregation that has been found to outshine all the existing network infrastructures.

REFERENCES

1. Sasipriya, S., and R. Vigneshram. An overview of cognitive radio in 5G wireless communications.In *2016 IEEE International Conference On Computational Intelligence and Computing Research (ICCIC)*, pp. 1–5. IEEE, 2016.
2. Rodriguez, Jonathan. *Fundamentals of 5G Mobile Networks*. John Wiley & Sons, 2015.
3. Liu, Chun-Nan. Trend, technology, and architecture of small cell in 5G era. In *2016 International Symposium on VLSI Design, Automation, and Test (VLSI-DAT)*, pp. 1–2. IEEE, 2016.
4. Kusaladharma, Sachitha, and Chintha Tellambura. An overview of cognitive radio networks. *Wiley Encyclopedia of Electrical and Electronics Engineering* 1 (1999): 1–17.
5. De, Parnika, and Shailendra Singh. Journey of mobile generation and cognitive radio technology in 5G. *International Journal of Mobile Network Communications & Telematics (IJMNCT)* 6 (2016): 9–17.
6. Badoi, Cornelia-Ionela, Neeli Prasad, Victor Croitoru, and Ramjee Prasad. "5G based on cognitive radio. *Wireless Personal Communications* 57, no. 3 (2011): 441–464.
7. Venkatesan, Mithra, A. V. Kulkarni, and D. Menon. Role of cognitive radio in 5G. *Helix* 9, no. 2 (2019), 4850–4854. doi:10.29042/2019-4850-4854.
8. Dadgarpour, A., M. S. Sorkherizi and A. A. Kishk. Wideband low-loss magnetoelectric dipole antenna for 5G wireless network with gain enhancement using meta lens and gap waveguide technology feeding. *IEEE Transactions on Antennas and Propagation* 64, no. 12 (2016), 5094–5101.
9. A. He, S. Srikanteswara, K. K. Bae, T. R. Newman, J. H. Reed, W. Hyy Tranter, M. Sajadieh, and M. Verhelst, System power consumption minimization for multichannel communications using cognitive radio. In *IEEE International Conference on Microwaves, Communications, Antennas and Electronics Systems (COMCAS)*, pp. 1–5, 2009.
10. Vahid, Seiamak, Rahim Tafazolli, and Marcin Filo. Small cells for 5G mobile networks. *Fundamentals of 5G Mobile Networks* (2015), 63–104.
11. Vidhya, R., and P. Karthik. Dynamic carrier aggregation in 5G network scenario. In *2015 International Conference on Computing and Network Communications (CoCoNet)*, pp. 936–940. IEEE, 2015.
12. Sharma, Shree Krishna, Mohammad Patwary, and Symeon Chatzinotas. Multiple access techniques for next-generation wireless: Recent advances and future perspectives. *EAI Endorsed Transactions on Wireless Spectrum* 2, no. 7 (2016): 1–10.
13. Kuo, Ping-Heng, and Alain Mourad. Millimeter wave for 5G mobile fronthaul and backhaul. In *2017 European Conference on Networks and Communications (EuCNC)*, pp. 1–5. IEEE, 2017.
14. Khan, Athar Ali, Mubashir Husain Rehmani, and Abderrezak Rachedi. When cognitive radio meets the Internet of Things? In *2016 International Wireless Communications and Mobile Computing Conference (IWCMC)*, pp. 469–474, IEEE, 2016.
15. D. Grace, J. Chen, T. Jiang, and P. D. Mitchell, Using cognitive radio to deliver green communications. In *4th IEEE International Conference on Cognitive Radio Oriented Wireless Networks and Communications (CROWNCOM)*, pp. 1–6, 2009.
16. A. He, S. Srikanteswara, J. H. Reed, X. Chen, W. H. Tranter, K. K. Bae, and M. Sajadieh, Minimizing energy consumption using cognitive radio. In *IEEE International Performance, Computing, and Communications Conference (IPCCC)*. IEEE, pp. 372–377, 2008.
17. Y. Saleem and M. H. Rehmani, Primary radio user activity models for cognitive radio networks: A survey. *Journal of Network and Computer Applications* 43 (2014), 1–16.
18. M. Höyhtyä, A. Mämmelä, M. Eskola, M. Matinmikko, J. Kalliovaara, J. Ojaniemi, J. Suutala, R. Ekman, R. Bacchus, and D. Roberson, Spectrum occupancy measurements: A survey and use of interference maps. *IEEE Communications Surveys and Tutorials* 18, no. 4 (2016), 2386–2414.
19. F. Akhtar, M. H. Rehmani, and M. Reisslein, White space: Definitional perspectives and their role in exploiting spectrum opportunities. *Telecommunications Policy* 40, no. 4 (2016), 319–331.

5 The Role of IoT in Smart Technologies

Ashish Bagwari
Women's Institute of Technology, Dehradun, India

Jyotshana Bagwari
Robotronix Engineering Tech Pvt. Ltd., Indore, India

Taniya Anand
Women Institute of Technology, Dehradun, India

Brijesh Kumar Chaurasia
Indian Institute of Information Technology (IIIT) Lucknow, Lucknow, India

R.P.S. Gangwar
Women Institute of Technology, Dehradun, India

Mohammad Kamrul Hasan
UniversitiKebangsaanMalaysia (UKM), Bangi, Malaysia

CONTENTS

DOI: 10.1201/9781003045809-6

5.1 INTRODUCTION

The Internet of Things (IoT) describes the interconnection of equipment having the ability to establish/ form a communication set-up. Devices with these capabilities are known as connected devices or smart devices. Electronic components such as software, actuators, network connectivity and sensors are embedded within objects like vehicles and buildings, enabling them to gather, trade, store and dissect information. The IoT along with its supporting technologies enables countless distinctive heterogeneous gadgets to be integrated and maintained to store, manage and prepare data for large-scale scientific advances. Open-source access is provided by IoT to classified subsets of data [1]. It is difficult to design and develop IoT-based smart systems because of the large variety of components that play key roles in their implementation, including sensors, actuators, network availability conventions and information, as well as their administration.

A good example of IoT innovation is the smart home, designed and maintained by IoT for comfort and convenience, with its services connecting the different appliances over a home network. To better monitor and control this smart home system, a wide variety of technology is used to equip entirely different devices. The smart home [1] includes smart appliances such as smart lighting circuits, heating and air-conditioning system, smart HDTVs, PCs, cellphones, smart refrigerator, security and CCTV systems. IoT enables smart home devices to interact together uninterrupted so that daily domestic activities can be automated without user intervention or controlled remotely more easily, safely and effectively. Users can control and manage them from anywhere on the planet with the help of the web. Put simply, a smart home is an IoT-enabled system of sensor networks that delivers all-time convenience to its owner for comfort, security and energy efficiency at a low cost.

A smart home provides various benefits in terms of energy efficiency by developing highly intelligent systems which extend the basic functionalities of domestic machines including programmed entry systems and light control. For home security there are IP-enabled devices such as cameras with smart features, safety advice or notifications, movement sensors and entry systems with smart features.

5.1.1 Components of the Smart Home

Smart home implementation requires the following resources [1, 3]:

- Interoperability of smart gadgets
- Central control box or gateway
- Smart home network
- Mobile and web applications for remote control.

Figure 5.1 shows a variety of interdependent devices that may be found in a smart home.

Interoperability of smart gadgets. A gadget with smartness is an electronic or electrical gadget consisting of actuators that are controlled by a microchip, microcontroller unit or micro-computer. Gadgets can be connected with peer gadgets or a network employing various remote network conventions, including Bluetooth Low Energy, Bluetooth, ZigBee, WiFi, near field communication, LoRaWan and Thread. They can thus operate somewhat intelligently and self-sufficiently and can share their sensors' information when required. Their information can also be permanently saved on the server hosted locally or CCB.

Central control box or gateway. The central control box (CCB) is a universally useful and appropriately programmed small PC. It acts as the entry point to the external interface, with the capacity to store a large amount of data, constant handling ability and web support, making it one of the main components of the smart home framework. Connecting the CCB or gateway

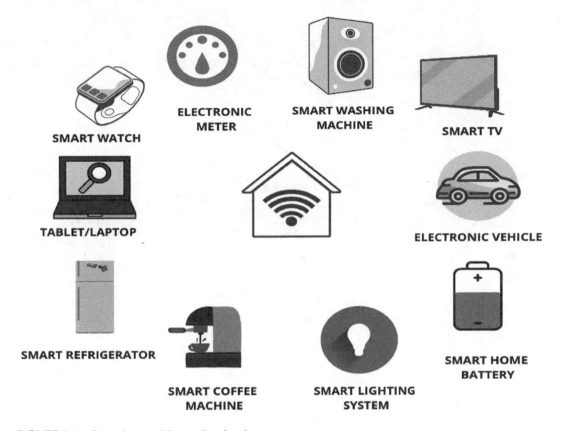

FIGURE 5.1   Smart home with associated gadgets.

to the external world using standard web conventions makes all the gadgets in the smart house available through mobile devices including smartphones and tablets, as well as via PCs from any location in the world. The gateway or CCB has various features including device set-up, identity management, access analysis, cloud administration, on-request device integration, administration and approval, and validation framework.

Smart Home Network. The inter-networking and correlation of various hubs or machines, the communication of network conventions and remote or wired connections is termed the smart home network. A mesh architecture and dynamic ability (because of portability of gadgets) has as its objective the communication of nodes inside the network as well as efficient provision of assistance to clients. Figure 5.2 is a home automation block diagram, showing how devices can communicate and be controlled over a home network. Mesh, star or peer-to-peer topologies along with central storage can be used to maintain this home network with wireless protocol. The diagram illustrates how sensors enabled on nodes or home machines can detect information from their internal state or the surroundings, which then acts as input for the controller device which is empowered with some logic. This controller can store the information in its internal storage system and then decide to activate the actuators to the device state.

Mobile and web applications. The thin applications of clients allow the availability of gadgets introduced in the smart house following verification from the entrance of the smart house or approval procedure from the service provider. It also combines web administration API with supporting protocols, allowing the client to design, oversee and access their IoT device. The CCB is used for notifying major corrections to devices and can also discover devices.

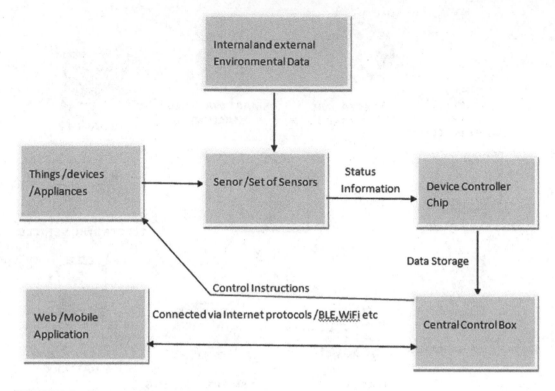

FIGURE 5.2 Home automation.

5.2 COMMUNICATION PROTOCOLS AND THEIR FEATURES

Network communication protocols are a key component of the IoT system and are responsible for secure and reliable transmission of data from one node to another or from source to destination. For optimum energy consumption, some buildings, farms or organizations currently have sensor-based applications. Areas including home automation, traffic lights, power grids and coal fields have devices based on sensors to implement systems for improved safety and security, convenience, overall cost-saving, and better transportation. Many security researchers and experts worry that IoT will make hacking much more common when consumers start purchasing dubious or second-hand smart gadgets for their homes. Detailed study of current communication protocols is fundamental for security and protection. All gadgets should have a compulsory underlying safety protocol that can be implemented naturally.

The main goal of communication protocols is the transmission of data from source to destination in a highly secure, financially efficient protocol based on wired or remote technology in a smart framework. In a wireless system, a protocol must be designed in such a way that the variety of gadgets in the market can communicate and exchange data effectively. The primary advantage of this system is that existing home appliances can be integrated into the control framework with little or no alteration. These protocols enable a level of compatibility with existing protocols so that the exchange of information or commands to the devices can be made from any location. A large amount of data is generated when devices use a system based on sensors, or during machine-to-machine communication; for example, a basic IP-empowered security camera with decent resolution can create more than 24GBof information every day. For such systems, the information is required to be hosted on huge cloud-empowered information-based systems, taking advantage of big data analytics.

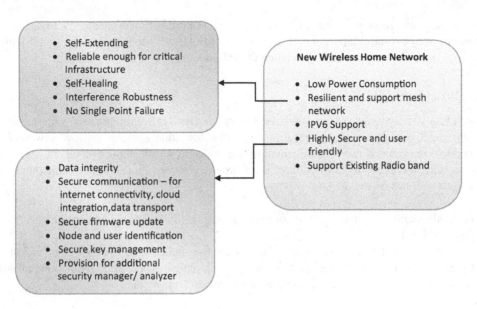

FIGURE 5.3   Desirable attributes for IoT communication protocols.

5.2.1   DESIRABLE ATTRIBUTES FOR IOT COMMUNICATION PROTOCOLS

Attributes desirable for IoT communication protocols are shown in Figure 5.3. The protocol may support a mesh/star topology network and may be able to increase the connectivity scope network for smart gadgets. If one node or device loses network availability and goes down, other nearby gadgets in the network must have adequate capacity to maintain network connectivity for data transmission. This is the basic benefit of a mesh topology.

Some gadgets that utilize the communication protocol for communication and data storage may be powerful systems including heating, ventilation and air conditioning, switches associated with smart lighting, water sprayers, automated curtains, internet protocol-empowered cameras, automatic washers, security systems (secured entrances, security cameras), electric attachment availability, TV and audio systems, automated window and garage closure. Smart home device services which will be supervised and maintained by communication protocol have the following features [4, 5]:

1. Interoperability with protocols that already exist.
2. On-call administration via client or things: efficient incorporation of cloud services.
3. Support authentication, management and preparation of information preparation: effective processing and storage of information.
4. Technology for security and protection: for point-to-point transmission and message-level security as well as efficient identification of devices, validation and reliability management to forestall local or remote access. Each layer of the protocol stack must provide high security and privacy.

5.3   BASICS OF PRIME IoT COMMUNICATION PROTOCOLS

Numerous transmission advances are widely used and well known in industrial and consumer IoT, for example, Bluetooth, Low Energy Bluetooth, WiFi, NFC, Sigfox, ZigBee and cellular networks. Emerging new networking IPV6-supporting protocols include Thread, an elective protocol for smart home applications. Whitespace TV technologies are also used for wide-area IoT-based use cases in major urban areas. Depending on the nature of the application, range, information transmission requirements, power requests, battery life, and security and protection are major factors that decide the choice of one or some combination of technologies to be used in IoT implementation. Significant smart home communication protocols include the following:

5.3.1 BLUETOOTH [6]

Bluetooth is a shorter-range (maximum over 100 m for class-1) communication remote protocol. It is a comparatively powerful, secure, low-cost technology that operates on low power. A time-division duplex scheme is used for information transmission in full-duplex mode. In other words, low-force radio waves are used to connect electronic devices using a wireless connection named Bluetooth. It operates on two levels:

1. For node identification and information transmission, it gives agreement among devices at the physical level.
2. It supports and uses next-level agreement, including gadget synchronizing and information transfer constraint over a time period. This procedure also ensures that the message transmitted and received between the communication parties is the same.

Bluetooth technology uses a combination of packet and circuit switching. To send/receive data Bluetooth applies a frequency-hopping spread spectrum approach which makes it very hard to track or capture transmissions. The Bluetooth protocol uses three transmission power classes. They are 1mW, 2.5mW and 100mW. This protocol uses a 48-bit unique hard-wired device address for unique identification of devices, which allows 2^{48}(2 to the power 48) devices. Bluetooth forms one or a set of piconets to communicate with devices and uses the master-and-slave node approach. Each piconet consists of up to eight active devices in which one is the 'master' and the remaining seven are 'slaves'. It also supports easy integration with TCP/IP for network connection and data accessibility. Bluetooth uses 2.45 GHz radio frequency and this frequency range is a globally available bandwidth used around the world for compatibility.

5.3.2 BLUETOOTH LOW ENERGY [7]

In standard Bluetooth, the interface is kept up irrespective of the information that transmission has occurred or not. BLE-enabled devices, utilized in IoT and developed typically for low-power applications, provide months of battery life because the sniff modes allow gadgets or devices to rest, greatly reducing power consumption. The highest transmit current is about 25 milliampere. The low energy of Bluetooth enables the innovation to interface the information to an internet service for processing. The BLE utilizes asynchronous connectionless MAC which contributes to low latency and extremely quick exchanges, generally 3 milliseconds from beginning to end.

5.3.3 ZIGBEE [8]

Quite similar to Bluetooth, ZigBee also has an enormous operation base installed. ZigBee PRO and ZigBee Remote Control (RF4CE) are the two variations that are IEEE802.15.4 protocol based. At a lower level of the protocol stack (physical and data link layers), it works on 2.4 GHz bandwidth, which requires relatively consistent information exchange at a lower information rate on a finite region only in a range of 100 meters, typical environments being a house or a small farmhouse. The basic advantages of this protocol are low power consumption, moderately higher protection, high scalability and sturdiness, with a large number of nodes in a WSN for M2M communication as well as IoT applications. The current version of ZigBee is 3.0 with a 10–100 meter range and 250 kilobits per second data rate.

5.3.4   Z-Wave [9]

A communication protocol based on a low-power radio frequency that is specially planned as well as produced for smart home gadgets like lamp regulators, smart light regulators, automatic gates and communication among numerous different gadgets. This communication protocol is proposed for genuine, restricted-range, small data packet communication based on low latency with an information transmit rate over 100 kilobits per second. In different countries, Z-wave works at 900 MHZ (ISM) frequency with 30 meters scope and a different frequency. From WiFi interference and other similar wireless protocols in the range of 2.4 GHz such as Bluetooth or ZigBee, this is impenetrable. Over 232 gadgets can be connected in a scalable full mesh network topology without the need for a coordinator node. It uses a set of simpler protocols which makes the development of applications fast and efficient, unlike other protocols. This protocol only supports Sigma Designs chips and has limited support for open standard chips compared to remote advancements like Bluetooth Low Energy, Sigfox, ZigBee, and so on.

5.3.5   IPv6LowPAN [10]

It is an IPv6-dependentinnovation with more nodes than other IoT application protocols such as Bluetooth Low Energy, Bluetooth, ZigBee, or Z-wave, with inherent security mechanisms. IPv6 Low-power Wireless Personal Area Network is a protocol for communication characterized by compression systems that are based on the compression of the header and encapsulated to increase the capacity of the packet payload. It enjoys bands of frequency in addition to the physical layer. It is interoperable with various communication networks formed with the help of sub-1GHz ISM, Ethernet, 802.15.4, and wireless fidelity. IPv6 offers 2128 unique addresses for physical devices or things in the world, enabling access through an internet connection. It is cost-effective, and specially designed and produced for home or building computerization, with a large number of nodes and a low-power RF wireless network.

This protocol makes use of IPv6 packets on networks that are dependent on IEEE802.15.4. It is executed on open internet protocol principles that include Web Socket, MQTT, COAP, Hypertext Transfer Protocol, TCP and UDP. It offers endwise addressable nodes and the functionality that associates the router from network to internet protocol. A self-repairing, strong and highly scalable mesh network which utilizes RFC6282 standard, with 2.4GHz bandwidth radiofrequency and sub-1GHz, 116 m range and information transmit rate of 10–50 kilobits per second are the additional services offered by this technology.

5.3.6   Thread [11]

Thread is a networking protocol that is based on IPv6 and especially designed for house mechanization technology. It is dependent on IPv6 over Low-power Wireless Personal Area Networks, and unlike ZigBee, Bluetooth Low Energy, Z-wave and Bluetooth protocols, it is not for IoT applications. It was originally developed as a supplement to WiFi, but it can be used to set up a home network for limited home mechanization. It was developed and released by the Thread Group in mid-2014 as a royalty-free protocol that depends on some standards including IPv6, IPv6 over Low-power Wireless Personal Area Networks, and IEEE802.15.4. It is suitable for an internet protocol-dependent network of IoT-empowered gadgets. Thread maintains a complex network of 250 nodes with a scope equal to 30 meters and a rate of information transmission equal to 250 Kilobits per second at 2.4 GHz (ISM) radiofrequency.

5.3.7   WiFi (Wireless Fidelity) [12]

A technology normally used in houses and businesses, utilizing IEEE802.11n standard. It provides faster information transfer equal to 1 Gbps (relying on 2.4 GHz as well as 5 GHz band channel frequencies) in the range of around 50 meters. There is a widespread existing framework that utilizes

WiFi as a communication protocol because of its faster data transfer rate and ability to deal with a larger amount of information than the protocols discussed above. It can be interpreted by ethernet, IP-enabled networks such asIPv6 over Low-power Wireless Personal Area Networks, and Thread, but it needs extra power. It may be one of the designs for automation of the smart home.

5.3.8 CELLULAR [13]

Cellular describes IoT applications that can be operated as well as controlled over an extended distance and can utilize the global cellular mobile communication protocols3G,4G or LTE for information exchange. It can transmit and receive a large amount of information at a speed of 35 to 170 Kilobits per second for General Packet Radio Service, 120 to 384 kilobits per second for Enhanced Data Rates for GSM Evolution, 384 Kilobits per second to 2 Megabits per second for Universal Mobile Telecommunications System, 600 Kilobits per second to 10Megabits per second for High-Speed Packet Access, 3 to 10 Megabits per second for Long Term Evolution and 20+ Megabits per second for 4G, which is expensive and uses a lot of power for data transmission. However, a sensor-based low-bandwidth-data project could be a very good choice to spread into a larger area and when a small amount of information has to be transmitted over the internet. A product using this protocol is the SparqEEscope of items, just like the first smallCELLv1.0 less expensive development board and a progression of safeguard interfacing sheets viable with the Arduino platforms and Raspberry Pi. Cellular protocols operate on various frequencies like 900/1800/1900/2100MHz and the distance range of this technology is 35km maximum for GSM and 200km maximum for HSPA.

5.3.9 NEAR FIELD COMMUNICATION (NFC) [14]

A communication protocol innovation that empowers safer basic two-way data exchange between electronic gadgets. It particularly operates for smartphones and NFC-enabled smart cards which allow consumers to share and access digital content, perform contactless payments and connect with electronic gadgets. In simple terms, it extends smart card payment capability as well as permitting gadgets to share relevant data at a distance of under 10cm. The standard use by NFC is ISO/IEC 18000-3, the range of frequency is 13.56MHz (ISM) and the information transmission rates: are 100–420Kilobitsper second.

5.3.10 SIGFOX [15]

It is a different wide-reach communication technology that is utilized by IoT-supporting gadgets. In rural areas, the scope for information transmission or control data transmission is 30 to 50 km and 3 to 10 km in metropolitan areas. The band used by this system is 900 MHz ISM bands. Sigfox supports M2M communication applications which require a small battery to run and can transmit a small amount of information. It is significantly less expensive than cellular and it has a higher range than WiFi. Its rate of data transfer is 10–1000 bps and it makes use of Ultra Narrow Band (UNB). For a single communication, it utilizes only 50 microwatts power, compared to cellular protocols which use 5000 microwatts power. A typical lifetime for a Sigfox-enabled device is around 20 years with the support of a 2.5 Ah powered battery.

5.3.11 LoRaWAN [16]

Similar to some extent to Sigfox, LoRaWAN is extensively used in the development of Wide Area Network (WAN) applications. Its normal reach is 5 km in metropolitan areas and 15 km in suburban areas. It is intended to execute low-power WANs suitable for low-cost mobile machine-to-machine communication, IoT, smart city, smart home and diverse modern applications. The information transfer speed ranges from 0.3 Kilobits per second to 50 Kilobits per second. It maintains millions of gadgets and devices in enormous networks.

5.4 RISKS WITH WIRELESS PROTOCOLS IN THE CONTEXT OF IoT

Most wireless protocols are very new technologies, but security and privacy issues must be very seriously considered. Data integrity and privacy risks are always inherent in any wireless technology because the communication path or medium is always open to each and every user, whether authentic users or intruders. Bluetooth and similar technologies like Bigzee, Z-wave, etc. use short-range radio which is vulnerable. For example, a hacker who has the frequency to connect to a PC within their range can use their own Bluetooth technology monitor and mouse to get access to the PC or device. And if the hacker's device connects to the victim's mobile phone by hacking the frequency, the victim will never be aware that someone has bugged or hacked their handset and everything will be at risk of unauthorized access. Communication protocols and applications therefore need to put extra effort into ensuring security and privacy to ensure the technology is safe for users. Above all, technology needs to address the following specific threats and issues [17, 18]:

- All the vulnerabilities, issues or threats that exist in a traditional wired network are available with wireless advances.
- Malicious nodes or units may gain unauthorized admission to an institution or IoT network through wireless connections, bypassing any gateway or firewall security.
- Confidential or private data that is unencrypted (or encrypted with poor cryptographic methodologies) and under transmission between two wireless nodes can be hacked and revealed.
- Denial-of-service assaults may be aimed at wireless connections or devices to increase their response time.
- The identity of valid users may be stolen by a malicious external entity mimicking them on internal or external corporate networks.
- Sensitive information may be damaged by an intruder during inappropriate synchronization or due to improper security majors.
- Malicious substances are capable of breaking valid users' protective measures and have the option to follow their developments.
- Handheld devices /IoT-based small devices in a smart home can be easily stolen and can reveal sensitive information.
- Information may be accessed from inappropriately configured gadgets and manipulated without detection.
- Malicious codes such as viruses or worms may damage data stored on the wireless device.
- Malicious entities may take advantage of wireless medium underuse and collaborate with other organizations to launch different kinds of assaults and conceal their activities.
- Interlopers, whether from the outside or inside, can have the option to access network management controls or tools and they can disable or interrupt tasks.
- Untrusted wireless network services can access an organization's network assets, as malicious entities might be introduced into the system by third-party resources.

5.5 CONCLUSION

This chapter has explored the IoT, smart homes and their underlying networks, IoT applications in smart homes, the essential structure required for implementation of a smart home and details of communication protocols, including security defects in the present communication protocols for IoT-enabled equipment. The chapter also explains communication protocols and related attributes required to achieve a protected, financially effective protocol intended for the smart home that can coordinate with activities such as existing protocol interoperability, gadget recognition, on-request administration or cloud administration incorporation with the client or with the device and its security and protection system.

REFERENCES

1. Saber Talari, Miadreza Shafie-Khah, Pierluigi Siano and Vincenzo Loia. "A Review of Smart Cities Based on the Internet of Things Concept." *Energies*, vol. 10, p. 421, 2017; doi:10.3390/en10040421. http://www.mdpi.com/journal/energies.
2. M. Razzaque, M. Milojevic-Jevric, A. Palade and S. Clarke, "Middleware for Internet of Things: A Survey." *IEEE Internet of Things Journal*, vol. 3, no. 1, pp. 70–95, 2016.
3. N. Komninos, E. Philippou and A. Pitsillides. "Survey in Smart Grid and Smart Home Security: Issues, Challenges and Countermeasures." *IEEE Communications Surveys & Tutorials*, vol. 16, no. 4, pp. 1933–1954, 2014.
4. John A. Stankovic, "Research Directions for the Internet of Things." *2014 IEEE. Personal use is permitted. For any other purposes, permission must be obtained from the IEEE by emailing pubs-permissions@ ieee.org*, IEEE.
5. J. Granjal, E. Monteiro and J. Sa Silva. "Security for the Internet of Things: A Survey of Existing Protocols and Open Research Issues." *IEEE Communications Surveys & Tutorials*, vol. 17, no. 3, pp. 1294–1312, 2015.
6. Sijan Shrestha, Cheikhoul Seck, Were Oyomno. "Security Protocols in Bluetooth Standard." *CT308A8800–Secured Communication*. December, 2007. http://edu.pegax.com/lib/exe/fetch. php?media=secc:bluetooth2007.pdf.
7. Matthew Bon. "A Basic Introduction to BLE Security." last updated on October 25, 2016. https://eewiki. net/display/Wireless/A+Basic+Introduction+to+BLE+Security. last accessed July 2017.
8. Pavel Ocenasek. "Towards Security Issues in ZigBee Architecture." M.J. Smith and G. Salvendy (Eds). *Human Interface, Part I, HCII 2009*, LNCS 5617, pp. 587–593, 2009. © Springer-VerlagBerlin Heidelberg, 2009.
9. Behrang Fouladi and Sahand Ghanoun. "Security Evaluation of the Z-Wave Wireless Protocol." *Sense Post UK Ltd*. http://neominds.org/download/zwave_wp.pdf. last accessed July 2017.
10. Jonas Olsson. "6LoWPAN Demystified." *System Applications Engineer Texas Instruments*. http://www. ti.com/lit/wp/swry013/swry013.pdf. last accessed July 2017.
11. Lucian Armas. "Thread Protocol: Enabling Secure Mesh Networks for Smart Home Devices." http:// www.tomshardware.com/news/thread-mesh-networking-protocol-homes,29556.html.
12. Vishal Kumkar, Akhil Tiwari, Pawan Tiwari, Ashish Gupta and Seema Shrawne. "Vulnerabilities of Wireless Security Protocols (WEP and WPA2)." *International Journal of Advanced Research in Computer Engineering & Technology*, vol. 1, no. 2, April2012. ISSN: 2278 – 1323.
13. Murat Oul and Selçuk Baktır. "Practical Attacks on Mobile Cellular Networks and Possible Countermeasures." *Future Internet*, 474–489, May2013, ISSN 1999-5903. doi:10.3390/fi5040474. www.mdpi.com/journal/futureinternet.
14. Security Concerns with NFC Technology. http://nearfieldcommunication.org/nfc-security.html. last accessed July 2017.
15. Make Things Come Alive in a Secure Way- Sigfox. February2017. https://www.sigfox.com/sites/default/ files/1701-SIGFOX-White_Paper_Security.pdf. last accessed July 2017.
16. Clemens Valens. "LoRaWANSecurity Vulnerabilities Exposed." October2016. https://www.elektorma-gazine.com/news/lorawan. last accessed July 2017.
17. Stephen Walsh, Jun Wan and Arran Sadlier. "Bluetooth Security." online. http://ctvr.tcd.ie/ undergrad/4ba2.05/group15/index.html. last accessed 12/7/2017.
18. Dimitris Geneiatakis, Ioannis Kounelis, Ricardo Neisse, Igor Nai-Fovino, Gary Steri and Gianmarco Baldini. European Commission, Joint Research Centre (JRC)Cyber and Digital Citizens' Security Unit Via Enrico Fermi 2749, 21027 Ispra, Italy. "Security and Privacy Issues for an IoT based Smart Home." *MIPRO 2017*, May 22–26, 2017, Opatija, Croatia.

Part II

Applied Scenarios of 5G and IoT

6 Realization of New Radio 5G-IoT Connectivity Using mmWave-Massive MIMO Technology

Priyanka Pateriya, Rakesh Singhai, and Piyush Shukla
Rajiv Gandhi Proudyogiki Vishwavidyalaya (RGPV), Bhopal, India

Jyoti Singhai
Maulana Azad National Institute of Technology (MANIT), Bhopal, India

CONTENTS

DOI: 10.1201/9781003045809-8

6.1 INTRODUCTION

The fifth-generation mobile communications network has attracted a great deal of attention and generated much research from global enterprises and research institutes. Many advanced high-speed applications, including smart cities, IoV, D2D, M2M and smart healthcare, will be part of the 5G-IoT revolution [1–4]. In the communications layer of 5G-IoT, 5G makes use of radio access technology (RAT) for IoT applications. 5G new radio (NR) is a Third Generation Partnership Project (3GPP) in next-generation wireless communications. 5G-NR technology is the part of RAT which is serviceable in the sub-6 GHz and 20–100 GHz (mmWave) range. 5G is essential to the IoT because of the need for a faster network with higher capacity that can respond to demands for high-speed connectivity, very low latency and ubiquitous coverage. The 5G spectrum extends the frequencies on which digital cellular technologies will communicate data, increasing the overall bandwidth (BW) of cellular networks and enabling additional devices to interconnect. 5G will facilitate remote regulation of more devices in applications where real-time network performance is crucial [5–7]. The three primary usage scenarios for 5G networks are stated by ITU and ETSI to be:

Enhanced mobile broadband. This aims to tackle the high user density data rate and traffic revolution together with. ITU standards will require 5G networks to support a high volume of user terminals (UTs) including IoT devices.

Massive machine-type communications. This standard is especially important when there is an immense number of connected devices. Devices are expected to have low power utilization and high data rates to satisfy ITU performance standards. ITU principles for mMTC concentrate on strengthening signaling protocols and messaging styles and techniques to reduce traffic congestion.

Ultra-reliable and low-latency communications (URLLC). Safety and mission-critical applications include IoT devices such as self-driving cars or medical devices. URLLC services are for applications that require very low latency and give real-time responses. The ITU guidelines for URLLC include additional service and content delivery, and a structured and systematic signaling protocol to address the restrictions of upgraded performance ongoing mobile systems for various kinds of applications.

Information theory suggests the following three primary concepts increase the network capacity of the wireless system [5, 6]:

- Ultra-dense networks.
- Wide bandwidth: proceeding towards higher frequencies will deliver the wide BW accessible to acquire higher capacity, specifically, millimeter-wave (mmWave) communications [8–11].
- High spectrum efficiency: by utilizing hundreds of transmitting and receiving antennas, mMIMO can enhance spectrum efficiency by comprehensively utilizing the acquirable space resources [12, 13].

The mMIMO system exploits an antenna array with an immense number of antennas (~ 64–128), which greatly improves the throughput of the system as well as simplifying the signal processing needed. It can assist a huge number of users, thus providing an enormous increment in network capacity. The mMIMO system is the main initiator in 5G, since it not only provides large directivity gains but also decreases the interference through beam shaping. Other than that, mmWave MIMO technology influences the wide spectrum in the mmWave band (30–300 GHz) to attain up to 100X growth in data rates over present ongoing systems, which encourages the realization of novel applications such as wireless HD, V2V/V2I communications and several others. Because of the smaller wavelength, large antenna arrays can be utilized in the mmWave communications method, proceeding to higher throughput through spatial multiplexing. The mmWave technology is especially attractive for mMIMO because it reduces the dimensions of the antenna array and tiny cell sizes are desirable, while the huge multiplexing and diversity gain presented by mMIMO helps to overcome the severe path loss of mmWave signals. Integrating mmWave and mMIMO technology to tackle the properties of large field throughput, coverage on appeal, and localized tiny cell hotspots as a consequence of mmWave technology – all supports the concept of mmWave-mMIMO, with wireless networking program [10–13].

Actualizing mmWave-mMIMO presents great challenges. For example, mMIMO needs systematic, methodical and effective algorithms for optimal decoding, and simple methods for channel estimation and precoding. The mmWave technology is challenging because of the large propagation loss at higher carrier frequencies and incremented signal obstruction along with sparse multipath propagation. Thus, hybrid RF-baseband processing seems to have become a popular architecture for mmWave-mMIMO systems, which require effective signal processing modules associated with channel estimation, beamforming, precoding and combining.

This chapter will discuss in detail the advancement of several concepts in mmWave-mMIMO technology for the IoT industry use cases enabled by 5G systems. This technology has enabled and empowered a huge number of applications directly intended to improve consumer outcomes (i.e., smart homes, healthcare, smart cities, industrial automation and entertainment, wearable devices, etc.). The chapter aims to:

- Highlight the waveform design methods for mmWave-mMIMO system in 5G-IoT networks
- Present the role of spatial multiplexing in the effective provisioning of communications networks
- Describe and classify the precoding/beamforming techniques for mmWave-mMIMO to meet the stringent 5G requirements
- Describe channel measurement and modeling and channel estimation methods for mmWave-mMIMO to meet the requirements of 5G IoT use cases
- Summarize the key technologies for realizing mmWave-mMIMO.

6.2 WAVEFORM DESIGN APPROACHES

This section reviews 5G waveform design approaches such as OFDM, FBMC, UFMC and GFDM to the provision of high-rate transmission [4, 14–18]. OFDM, FBMC and UFMC are multicarrier systems that use large numbers of subcarriers as parallel narrow band carriers rather than a single wideband carrier to transmit message signals. FBMC and GFDM are subcarrier-wise filtered multicarrier schemes, whereas UFMC is a subband-wise filtered multicarrier scheme.

6.2.1 OFDM

OFDM is used in LTE systems and is an essential 5G technology contender because it has the potential to resist multipath distortion and effortlessly implement FFT and IFFT blocks. OFDM can deal with frequency-selective channels. The data stream is partitioned into parallel streams, where each data stream is modulated with a set of narrow subcarriers whose BW has to be short compared to the coherence BW of the channel in such a way that every single subcarrier undergoes a flat fading channel that can be equalized in the frequency dimension with the help of mathematical operations. An OFDM system uses the spectrum efficiently because of the orthogonally overlapped subcarriers. The benefits of OFDM waveforms are:

- The orthogonal property of subcarriers decreases the consequences of intercell interference and enables them to adjust phase and frequency distortions to permit OFDM to perform with MIMO technology efficiently.
- OFDM can be implemented easily.

- Low complexity because separate subcarriers are allocated to separate users.
- Excessive data-rate transmission can be attained by wide BW which in response realizes a frequency-selective channel. The OFDM technique helps convert these channels to a parallel flat fading channel for examining signal components.

However, OFDM has a number of disadvantages:

- High PAPR because of the arbitrary insertion of subcarriers in the time dimension. The variance of the output power also increases in line with the number of subcarriers.
- Highly sensitive to synchronization errors. The solution is to utilize the CP, although this creates a guard interval, which reduces spectral efficiency.

6.2.2   FBMC

FBMC is a multicarrier system that provides better frequency-domain localization and utilizes structured pulse-shaping filters, which are practicable at the subcarrier level, and facilitates modifications to channel situations, thus helping reduce signaling overheads to improve latency and strengthen user services in 5G-IoT wireless connectivity. There are several implementation methods for FBMC, including CMT, SMT and FMT. SMT and CMT are significant FBMC schemes for maximizing bandwidth efficiency, SMT utilizes offset quadrature amplitude modulation to transmit message symbols, and CMT utilizes pulse amplitude modulation. Both schemes allow the overlapping of adjacent bands to achieve efficient bandwidth. SMT is also called OQAM-FBMC, which is the basis of 5G waveform analysis because of its capability to tackle interference. The terms 'in phase' and 'quadrature phase' are used in time and frequency dimensions in OQAM signaling, and therefore orthogonality is preserved within the real and imaginary domains individually. Furthermore, its capability to attain network synchronization is essential for 5G-IoT systems

FBMC subject to mMIMO systems facilitates self-equalization, which presents several advantages, such as low latency and complexity, increases BW efficiency, and reduces PAPR. Pilot contamination is a big challenge for channel estimation in the mmWave-mMIMO system.

6.2.3   GFDM

GFDM is a non-orthogonal transmission scheme with non-orthogonal filters. The basic concept of GFDM is block frame transmission, which comprises M time slots and K subcarriers. Its flexibility and block structure helps in the fulfillment of URLLC of 5G-IoT systems. GFDM is different from FBMC in the sense that the filters for pulse shaping in GFDM are circularly convoluted over the transmitted signals. The transmitted signals are processed block-wise, and the CP is inserted into this block. The number of subcarriers, filters and sub-symbols can be adapted for different channel conditions, making GFDM a flexible waveform. GFDM signaling can also be actualized using FFTs and IFFTs blocks. GFDM is appropriate for wireless communications; it is highly immune to synchronization errors because it shares the frequency-localized characteristics with OQAM-FBMC. This method requires SIC algorithms at the receiving side. The block-wise transmission results in increased latency that marks it impractical for mission-critical applications, a disadvantage that gives rise to sub-band filtering.

6.2.4   UFMC

UFMC is a non-orthogonal waveform, introduced by 5G-NOW groups. UFMC utilizes subband-wise filtering to decrease out-of-band emissions. This filtering reconciles the subcarrier-wise filtering and band filtering. The filters are lessened compared with the FBMC, where the size of subcarrier-wise filtering is larger than the symbol duration. Another difference is that FBMC utilizes

TABLE 6.1

Advantages and Disadvantages of Various 5G Waveform Schemes

Waveform	Advantages	Disadvantages
OFDM	Low implementation complexity Flexible frequency localization Easy MIMO integration Flexible filtering granularity Shorter filter length	High PARP Performance level is low for high mobility services
FBMC	Low OOBE Good spectral efficiency Poor performance for huge mobility services Efficient for asynchronous transmission	Pilot contamination High power consumption
GFDM	Flexibility in design Better frequency localization Decreased PARP	High latency due to block processing
UFMC	Good frequency localization Shorter filter design Flexible filtering granularity Compatible with mMIMO	Less immune to ISI Receiver complexity is high

filters on every single subcarrier, whereas UFMC implements filtering on a few subsets of the sub-carriers, and provides superior spectral efficiency to OFDM. It supports the grouping of the subcar-riers. UFMC decreases the filter length, making it compatible with low-latency services and more robust against inter-carrier interference. The disadvantage of UFMC is the high level of complexity caused by challenging and difficult filtering activities for the design of practical communications systems (Table 6.1).

6.3   SPATIAL MULTIPLEXING

Spatial multiplexing (SM) is a transmission scheme in the wireless network for transmitting inde-pendent and distinct encoded data streams, referred to as 'streams', so that the spatial domain is multiplexed and utilized more than once. This transmission technique demultiplexes the input data stream into N independent streams through the S/P converter, and every single stream transmits from an individual transmitting antenna. Therefore, the throughput is N symbols per channel use for a MIMO channel with N_t transmitting antennas. This throughput increases at the price of low diver-sity gain in contrast to STC. SM increases throughput by transmitting message data onto parallel streams. It is therefore a superior option for high data-rate systems operating at high SNR, whereas STC is a superior option for transmitting message signals at reasonably low data rate and low SNR [19, 20].

The mmWave systems are appropriate for employing an immense number of antennas oper-ating at very high frequency bands. In mmWave channels, SM provides the multiplexing gains that enhance transmission throughput by partitioning the transmitted data stream into several data streams, which are transmitted concurrently in parallel on the same frequency band via the inde-pendent antenna. The multiplexing gain can be attained by manipulating the spatial variation of the channel response in the specific number of transmitting-receiving antenna sets.

In the mmWave-mMIMO system, SM can be realized through the SVD method. SM makes use of the advantage of the dissimilarities in the channel parameters among the transmit and receive antenna pairs to provide several independent streams amongst the transmit and receive antennas. SVD partitions the total transmission power among the multiple transmitting streams and calculates the SNR and throughput for each parallel data stream, which are then added together to calculate the aggregate throughput from all the multiple transmitting streams. Thus, the multiplexing gain of SM = min (N_t, N_r).

The primary approach of the SM coding framework is the V-BLAST scheme, which transmits the NL symbols over L symbol periods from the N_t transmitting antennas. SIC is a popular detection scheme. SIC detects N received symbols in sequence and removes interference through the hard-decision approach. SIC performs the three significant operations known as ordering, nulling and cancellation.

The ordering operation selects the order of detection of the symbols at each time step. The nulling step can be realized through ZF or MMSE linear detectors to provide the best estimate of each transmitting data stream in the presence of interference and noise. The cancellation operation enhances the performance of subsequent nulling loops by eliminating the interference produced by the most recently decoded symbol. This step depends on the hard-decision or quantized symbol. The diversity gain attainable through V-BLAST is M, because M independent faded copies of every transmitted symbol are observed by the receiver.

6.4 PRECODING

Precoding is a signal processing technique that uses the CSI at the transmitting end to maximize the link performance of the communications system. Downlink transmission performance is dependent on CSI at the transmitter and the corresponding precoding technique used. Where the CSI is well known at the transmitting end, precoding can reduce interference and enhance the achievable sum rates. The multiple-antenna system with the appropriate precoding method increases the spectral efficiency and energy efficiency [13, 21–28] of the communications system. There are three significant precoding techniques, which are discussed in the next sections.

6.4.1 DIGITAL PRECODING

This precoding technique is utilized in MIMO systems that operate on low frequencies. The idea is to regulate the phase and amplitude of the message signals to nullify interference in advance. This technique can be categorized as linear precoding (LP) or nonlinear precoding (NLP). In LP, the transmitted signals are produced by the linear combination of the message signals. MF and ZF precoder techniques are linear precoding schemes. In NLP, the transmitted signals are produced nonlinearly. Block diagonalization (BD) is a nonlinear precoding scheme. Digital precoding can be further sub-divided into SU and MU digital precoding.

6.4.1.1 SU Digital Precoding

Figure 6.1 illustrates a digital precoder (D) in the SU mmWave-mMIMO, where the BS exploits N_t number of transmitting antennas and a user with N_r number of receiving antennas with the condition that N_r should be less than N_t and the number of RF chains are equal to N_t. The BS is equipped with $N_t \times N_r$ digital precoder D. Assuming that the channel matrix (H) of dimension $N_r \times N_t$ with normalized power $E\left(\left\|H\right\|_F^2\right) = Nt.Nr$ is known at the BS to facilitate the precoding technique. The transmitted signal (X) can be represented as:

$$X = Ds \qquad\qquad (6.1)$$

where s is the $N_r \times 1$ message vector prior to precoding with normalized power given as $E(ss^H) = (1/N_r)I_{Nt}$.

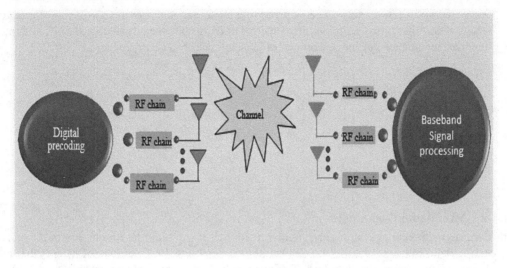

Base Station   User-Terminal

FIGURE 6.1 The architecture of the SU mmWave-mMIMO system with digital precoding.

D needs to fulfill the overall transmit power constraint condition $\|D\|_F^2 = \mathrm{tr}(DD^H) = N_r$. The received signal (y) of dimension $N_r \times 1$ can be represented as:

$$y = \sqrt{\rho}\, HDs + n \tag{6.2}$$

where ρ = average received power, and n= AWGN vector.

- *Matched filter precoding* is a simple linear-digital precoding technique. This precoding technique increases the SNR at the receiver. It can be expressed as

$$D = \sqrt{\frac{N_r}{tr\left(FF^H\right)}}\, F, \tag{6.3}$$

where $F = H^H$

This precoder causes drastic interference between the transmitted data streams. The solution to this problem is to utilize the zero-forcing precoder.

- *Zero-forcing precoding* can completely remove interference amongst the transmitted data streams. To satisfy the overall transmitting power constraint, the ZF technique can increase the noise power, which causes performance loss. ZF precoder can be expressed as

$$D = \sqrt{\frac{N_r}{tr\left(FF^H\right)}}\, F, \tag{6.4}$$

where $F = H^H(HH^H)^{-1}$

- *Wiener filter precoding.* The MF precoder enhances the SNR at the receiver at the cost of increased interference, whereas the ZF precoder decreases the interference at the cost of increased noise power, so we need a precoding technique that presents a better performance. The Wiener filter precoder provides a superior trade-off between the received SNR and interferences to enhance system performance. This precoder can be expressed as

$$D = \sqrt{\frac{N_r}{tr\left(FF^H\right)}}\, F, \qquad (6.5)$$

where, $F = H^H(HH^H + \frac{\sigma_{nN_r}^2}{\rho} I)^{-1}$

6.4.1.2 MU Digital Precoding

In the MU digital precoder system for the mmWave-mMIMO system, the BS comprises N_{BS} antennas and N_{BS} radio frequency chains to be in communication with the U number of MSs at the same time. Every MS employs N_{MS} antennas. The numerous data streams for communication is $N_{MS}U$ with the condition that $N_{MS}U \leq N_{BS}$.

In the downlink transmission mode, the BS is equipped with a precoder of dimensions $N_{BS} \times N_{MS}U$ denoted by $D = [D1, D2,........, D_U]$ for the uth user terminal. The MU digital precoder should fulfill the overall transmit power constraint $\|D\|_F = N_{MS}$. The signal vector r_U received by the uth user can be written as:

$$r_U = H_U \sum_{n=1}^{U} D_n S_n + n_U, \qquad (6.6)$$

$r_U = H_U$ where S_n is the message signal of dimension $N_{MS} \times 1$ prior to precoding,

H_U is the mmWave-mMIMO channel matrix amongst the BS and the uth mobile station. This matrix is of dimension $N_{MS} \times N_{BS}$,

n_U is an i.i.d AWGN vector at the uth user terminal,

It can be seen from Equation (6.6) that the components $H_U D_n S_n$ for $n \neq u$ are interferences to the uth mobile station. Thus, a precoder is required to design all D_n that must lie in the null space of $\overline{H_U}$, i.e., $\overline{H_U}.D_n = 0$ to remove the interference components. This type of precoding technique is known as block diagonalization. The channel matrix $\overline{H_U}$ can be expressed as:

$$\overline{H_U} = \left[H_1^H,.........,H_{U-1}^H, H_{U+1}^H,.........,H_U^H \right] H \qquad (6.7)$$

The digital precoding technique regulates the phase and amplitude of transmitted signals to improve system performance. The disadvantages of this precoder are its huge energy consumption and hardware cost, because the number of RF chains is directly proportional to N_t making it challenging for mmWave-mMIMO systems.

6.4.2 ANALOG BEAMFORMING

Analog beamforming uses a single radio frequency chain to transmit a single data stream. This technique regulates the phase of the message signal to attain the maximum antenna array gain and SNR. Popular beamforming techniques are beam steering and beam training.

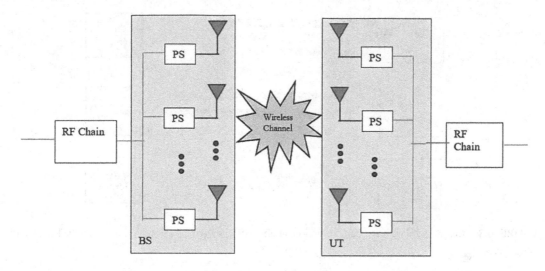

FIGURE 6.2 The architecture of the SU mmWave-mMIMO system with analog beamforming.

6.4.2.1 Beam Steering

Figure 6.2 illustrates the analog beamformer (A) in the SU mmWave-mMIMO system,, where the BS possesses N_t antennas and a single RF chain for transmitting the message stream to the receiver and employs N_r receiving antennas and a single radio frequency chain. The CSI is known at the transmitting and receiving side and defines an analog beamforming vector f of dimension $N_t \times 1$ at the BS and an analog combining vector w of dimension $N_r \times 1$ at the UT. The beamforming is combined with vectors to increase the SNR at the receiver; these vectors can be represented as:

$$\left(w^{opt}, f^{opt} \right) = \arg\max \left| w^H H f \right|^2 \tag{6.8}$$

For mmWave-mMIMO systems, the beamforming and combining vector can be limited to the array response vectors. Selecting $f = a_t\left(\varnothing_{k^*}^t, \theta_{k^*}^t\right)$ and $w = a_r\left(\varnothing_{k^*}^r, \theta_{k^*}^r\right)$, where $\left(\varnothing_{k^*}^t, \theta_{k^*}^t\right)$ and $\left(\varnothing_{k^*}^r, \theta_{k^*}^r\right)$ are the quantified AOD and AOA, accordingly and $k^* = \arg\max_l |\alpha_l|^2$ to steer the beam in the right direction to enhance the performance of the mmWave-mMIMO system.

6.4.2.2 Beam Training

In practical systems, the availability of CSI at both ends is challenging. The solution to this problem is the introduction of the beam training technique to achieve superior beamforming samples/vectors deprived of the knowledge of CSI. In the beam training technique, the BS and the UT should associate to establish the best beamforming and combining set from predefined codebooks. The codebook can be modeled according to the beam steering scheme. The popular training algorithm thoroughly searches all probable beamformer |f| and combiner |w| pairs to increase the effective SNR. Despite that, in mmWave-mMIMO systems, the large N_t and N_r and beamforming gain condition will greatly increase the size of the codebook, which is a challenge.

The solution is to enhance the resolution of codebooks f1, f2, …, fK (w1, w2, …, wK). Now, the BS and the UT together guide the beams at the first level (lowest resolution codebook f1) by transmitting the training data, which comprises three steps:

- The BS transmits the training data through f1 to the UT, which evaluate the best combiner vector as shown in Figure 6.3(a).

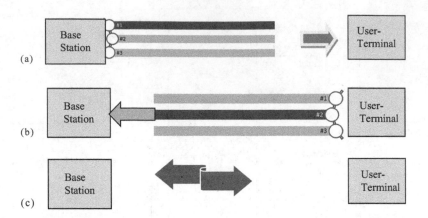

FIGURE 6.3 Beam training: (a) beam selection at the BS end, (b) beam selection at the UT end and (c) feedback step.

- The UT and the BS exchange their responsibilities and evaluate the superior beamformer vector as shown in Figure 6.3(b) and then BS and UT feed the chosen beam information back to each other as shown in Figure 6.3(c).
- These steps are repeated in the selected beam till the higher resolution codebook to decrease overheads, instead of an exhaustive search.

The analog beamformer needs a single RF chain and thus is easier to implement than digital precoding. But this technique suffers from performance loss because it regulates only the transmitted signal phase. Significantly, this scheme is easy to use in SU systems and difficult to use in an MU-multistream system.

6.4.3 HYBRID PRECODING

Digital precoding and analog beamforming suffer a few challenges when the number of antennas increases, therefore a new precoding technique known as 'hybrid precoding' is required to deal with the mmWave-mMIMO system [14, 29–33]. The hybrid precoding scheme uses the digital precoder (D) and analog beamformer (A) effectively. This precoding scheme employs a small dimension D to remove interferences between the transmitted data streams in advance and a large dimension A to maximize the antenna array gain. Hybrid precoding can be categorized into two groups on the basis of structural implementation:

- *Fully connected architecture*. In this, every RF component is linked to all antennas at the BS through phase shifters. Spatially sparse hybrid precoding is commonly used in connected architecture.
- *Sub connected architecture*. In this, every RF component is linked to the antennas subset at the BS. SIC is a commonly used precoding architecture.

Furthermore, this precoding can also be categorized as SU and MU. Furthermore, this precoding can also be categorized based on SU and MU. Both the schemes can implement in the form of fully connected architectures as well as sub connected architectures.

6.4.3.1 SU Hybrid Precoding

Figure 6.4 shows the SU mmWave-mMIMO system with hybrid precoding. In this system, the BS is furnished with N_t antennas for transmitting N_s data streams simultaneously to a UT equipped with N_r antennas. Multi-stream transmission is enabled if the BS is employed with N_t^{RF} radio frequency

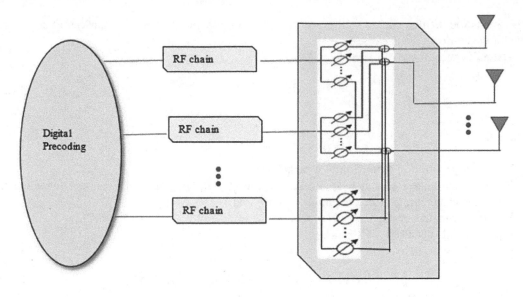

FIGURE 6.4 Single-user mMIMO system utilizing fully connected hybrid precoding.

chains so that $N_s \le N_t^{RF} \le N_t$. The BS employs a digital precoder D of dimension $N_t^{RF} \times N_s$ utilizing its N_t^{RF} radio frequency chains, followed by an analog beamformer A of dimension $N_t \times N_t^{RF}$ utilizing analog circuits like PSs. The X can be expressed as:

$$X = ADs \qquad (6.9)$$

where s is the message vector of dimension $N_s \times 1$ prior to precoding with normalized power of $E(ss^H) = (1/N_s)I_{N_s}$. The y of dimension $N_r \times 1$ can be expressed as:

$$y = \sqrt{\rho}\,HADs + n \qquad (6.10)$$

In this realistic system, the receiver can determine the CSI through training and afterwards feed the CSI back to the transmitter.

The spatially sparse precoding fully connected architecture connects each RF chain to each transmit antenna at the BS through PSs as presented in Figure 6.4. The analog PSs are utilized to implement an analog beamformer A, where every element of an analog beamformer has the same amplitude $= 1/\sqrt{N_t}$ but a distinct phase. The total transmits power constraint is imposed by normalizing D to fulfill the condition that $\|AD\|_F^2 = N_s$. The A,D is designed in such a manner that it maximizes the sum rate R(A,D) over the mmWave channel environment.

$$R(A,D) = \log_2\left(\left|I + \frac{\rho}{N_s\sigma_n^2}HADD^H A^H H^H\right|\right) \qquad (6.11)$$

The optimum value of A,D that solves the optimization problem to achieve the huge sum rate is expressed as:

$$\left(A^{opt}, D^{opt}\right) = \arg\max_{A,D} R(A,D) \qquad (6.12)$$

$$s,t\ A \in F$$

$$\|AD\|_F^2 = N_s, \qquad (6.13)$$

where F is the set that contains all practicable As, i.e., the set of $N_t \times N_t^{RF}$ matrices i^{th} constant-magnitude entries.

The SVD decomposes the channel matrix as

$$H = U \Sigma V^H \tag{6.14}$$

where $\Sigma = \begin{bmatrix} \Sigma_1 & 0 \\ 0 & \Sigma_2 \end{bmatrix}$, $V = \begin{bmatrix} V1, V2 \end{bmatrix}$

U = unitary matrix of dimension $N_r \times$ rank(H) and

Σ = diagonal matrix of dimension rank(H) $\times$ rank(H), the diagonal elements are the singular values of the channel matrix in descending manner, and

V = unitary matrix of dimension $N_t \times$ rank(H).

Equation (6.11) can be rewritten as:

$$R(A,D) = \log_2\left(\left\|I + \frac{\rho}{N_s\sigma_n^2}\Sigma_2\, VHAD\, D^H A^H V\right\|\right) \tag{6.15}$$

Note that the Σ_1 is of dimensionality $N_s \times N_s$ and V_1 is of dimensionality $N_t \times N_s$. Observing that the dimension of V_1 is similar to the optimal unconstrained precoder (P_{opt}), thus $P_{opt} = V_1$. However, this optimum precoder P_{opt} cannot be realized utilizing this architecture. Thus, we need to model a realistic AD near to the optimum unconstrained precoder V_1, to acquire optimum performance. Algebraically, this 'closeness' is characterized by the given two approximations:

- The eigenvalues of the matrix $\left[I - V_1^H ADD^H A^H V_1\right]$ are small. For the mmWave-mMIMO precoding, this can be indicated as $V_1^H AD \approx I$.

- The singular values of the matrix $V_2^H AD$ are small; alternatively $V_1^H AD \approx 0$.

Based on the above two assumptions, the R(A,D) can be written as:

$$R(A,D) = \log_2\left(\left\|I + \frac{\rho}{N_s\sigma_n^2}\Sigma_1^2\right\|\right) - \left(N_s - \left\|V_1^H AD\right\|_F^2\right) \tag{6.16}$$

The above equation shows that the relation between the R(A,D) and the modeled precoder AD is calculated by the component $\|V_1^H AD\|_F^2$. The term $tr(V_1^H AD)$ is inversely proportional to $\|P_{opt} - AD\|_F$. Approximately, the R(A,D) optimization problem is written as:

$$\left(A^{opt}, D^{opt}\right) = \arg\max_{A,D} \|P_{opt} - AD\|_F \tag{6.17}$$

$$s, t\ A \in F$$

$$\|AD\|_F^2 = N_s,$$

This precoding scheme formulates R(A,D) through a sparse approximation algorithm. This algorithm inputs an optimal unconstrained precoder corresponding with the linear aggregate of beam steerage vectors that can be realizable through analogy circuit together with a digital precoder at the BS to achieve the optimum hybrid precoder.

6.4.3.2 MU Hybrid Precoding

The MU mmWave-mMIMO system is presented in Figure 6.5. The BS is equipped with N_{BS} antennas and N_{RF} RF chains for communicating with U number of mobile stations at the same time with the condition that $N_{RF} \leq N_{BS}$. Each MS employs N_{MS} antennas and a single RF chain. This section focuses on the case where the BS is communicating with every MS through a single data stream. Thus the numerous data-communication streams are $N_S = U \times N_{RF}$. This assumes that the BS utilizes the U out of the N_{RF} RF chains to communicate with U MSs to attain the high spatial multiplexing gain.

In downlink data transmission, the BS employs the fully connected architecture equipped with a digital precoder denoted by D = [d1, d2,…,d_U] followed by an analog precoder denoted by A = [a_1, a_2,…a_U] of size $N_{BS} \times U$. The transmitted signal can be represented as

$$X = ADs$$

where $s = [s_1, s_2, \dots s_U]^T$ is the message signal vector of dimension U×1 prior to precoding with normalized power $E(ss^H) = (\rho/U)I_U$ and

ρ = average transmitted power.

Considering the narrowband block-fading channel model and also assuming equal power allocation between different MSs data streams because all the elements of A have the equal amplitude of N_{BS}^{-1}. The overall transmitted power constraint is imposed by normalizing D to fulfill $\|AD\|_F^2 = U$. The received signal vector r_u examined by the uth MS can be expressed as:

$$r_u = H_u \sum_{n=1}^{U} Ad_n s_n + n_u, \tag{6.18}$$

where H_u = mmWave-mMIMO channel matrix between the BS and the uth MS of dimension $N_{MS} \times N_{BS}$

n_u is an i.i.d., AWGN vector

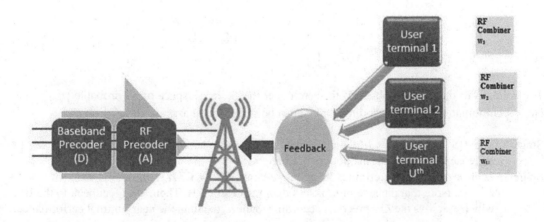

Base Station   **User-Terminal**

FIGURE 6.5 MU mMIMO system architecture using hybrid precoding technique.

The analog combiner (w_u) is utilized at the MS to reduce the hardware requirement. w_u has the same constraints as the A, i.e., all the elements of w_u possess a similar amplitude N_{MS}^{-1} but different phases. At the uth MS, w_u is utilized to merge the received signal r_u as shown in Figure 6.5.

$$y_u = w_u^H r_u = w_u^H H_u \sum_{n=1}^{U} A d_n s_n + w_u^H n_u \qquad (6.19)$$

6.4.3.2.1 Two-Stage Hybrid Precoding

The goal of this method is to augment the data transmission rate (R) by modeling the A,D at the BS, and $\{w_u\}_{u=1}^{U}$ at the MSs. This method partitions the design of hybrid precoding into two stages. First, A and w_u are modeled jointly to maximize signal power at user terminals. Second, D is modeled to reduce multi-user interference. D requires to be modeled according to the analog precoding/combining vectors. The sum rate is given by:

$$R = \sum_{u=1}^{U} R_u, \qquad (6.20)$$

where $R_u = \log_2 (1 + \dfrac{\dfrac{P}{U}\left|w_u^H H_u A d_u\right|^2}{\dfrac{P}{U}\sum_{n \neq u}\left|w_u^H H_u A d_u\right|^2 + \sigma_n^2})$, is the data transmission rate attained by the uth MS.

Now the precoding modeling problem can be formulated to determine A^{opt}, D^{opt}, and $\{w_u^{opt}\}_{u=1}^{U}$ which solves the problem below

$$\left\{A^{opt}, D^{opt}, \{w_u^{opt}\}_{u=1}^{U}\right\} = \arg\max \sum_{u=1}^{U} \log_2 \left(1 + \dfrac{\dfrac{P}{U}\left|w_u^H H_u A d_u\right|^2}{\dfrac{P}{U}\sum_{n \neq u}\left|w_u^H H_u A d_u\right|^2 + \sigma_n^2}\right) \qquad (6.21)$$

$$\left.\begin{array}{l} s,t\, a_u \in F, \\ w_u \in W, \\ \left\|AD\right\|_F^2 = U \end{array}\right\} \quad \text{for, } u = 1, 2, \ldots., U \qquad (6.22)$$

The solution to the above problem is to search over the $F^U \times W^U$ space of all probable $\{a_u\}_{u=1}^{U}$ and $\{w_u\}_{u=1}^{U}$ combinations. This precoding method can be summarized as:

Stage 1. The BS and each uth MS models the analog precoder/combiner vectors a_u and w_u, to increase the desired signal power for the uth MS.

Stage 2. Each uth MS first determines its effective channel $\bar{h}_u = w_u^H H_u A$ of size U × 1, for u = 1, 2,…,U. Note that the size of $\bar{h}_u$ is less than the original H. Then, $\bar{h}_u$ is feedback to the BS, which employs the ZF- precoder according to the $\bar{h}_u$ to attain the near-optimal performance.

6.5 CHANNEL MEASUREMENT AND MODELING

The mmWave-mMIMO channel is an ultra-broadband channel with a large BW and SM ability to increase internet access and increases the cell and UTs outcomes. Channel analysis and development can be done through mathematical analysis and field management of the channel

[13, 34, 35]. The channel modeling for the multiple antennas system operating at mmWave is discussed below.

6.5.1 CHANNEL MEASUREMENT

This technique is utilized for examining wireless channel properties. Channel sounding is a popular field measurement technique for mmWave frequency bands as it is the instrument that provides an efficient impulse response or angular properties of the channel. The channel sounder is significant for mmWave-mMIMO channel modeling, test-bed, simulation and validation as it provides precise calibration, synchronism, resolution and susceptibility at reasonable system cost. The measurement requirements of the sounding approach depend upon the frequency and BW of the channel, the channel detection unit for multiple antennas, and whether the channel is static or dynamic to maintain the speed of the sounder.

The other fundamental requirements of the channel sounder for mmWave frequency bands are:

- RF transceiver front-end devices
- Signal generation and acquisition for the high performance with wideband DACs and ADCs
- Effective data storage
- Precise calibration and synchronization to fulfill the increasing demand of high path delay resolution and high susceptivity in the interchannel delay estimation.

An efficient sounder for the mmWave channel is one that fulfills the above requirements. Wideband signal correlation delivers high processing speed and receives a complex signal that can be distributed in angular manner in multiple-antenna systems. Because of these advantages, the wideband signal correlation approach is appropriate for mmWave-mMIMO sounding.

6.5.2 CHANNEL MODELING

Channel modeling develops the mathematical and simulation frameworks for the comparison and validation of the empirical data from field tests to evaluate the wireless network channel parameters. The 3D geometry-based SCM method is the most popular method presented by the 3GPP.

SCM models the positions of the BS utilizing the PPP-based abstraction model. SCM models the realistic channel characteristics of mmWave propagation through the three-state (LOS, NLOS and outage) link model. In communications networks, system-level performance is evaluated by numerical simulations and mathematical analyses to ensure high confidence levels and incorporate all network parameters and data sets that can impact performance. SCM is a flexible mathematical tool for analyzing future cellular networks in order to deal with these challenges.

6.6 CHANNEL ESTIMATION

Channel estimation (CE) is a significant task for wireless communications systems evaluating the CSI. CSI characterizes the way a signal is broadcast from the transmitting terminal to the receiving terminal and includes the effect of scatterings, fading interference and power decay with distance. In practice, channel parameters change with time because of environmental changes, obstacles, and so on, that scatter the signal. Channel information needs to be accessible at the transmitter for precoding on the downlink communication mode to achieve optimum performance. The CSI is achieved through uplink pilots with channel reciprocity in TDD-based communications systems or through feedback to the BSs by the UT in the FDD-based communications system [13, 29–33, 36–38].

TDD protocol utilizes the channel reciprocity approach for MIMO systems, where uplink pilot signals are utilized to evaluate the channel parameters and subsequently downlink precoding and data transmission. For mmWave-mMIMO systems, the channel information at the BS is necessary for the realization of SM and is crucial for high system performance. The utilization of uplink pilots relies upon the number of receiving antennas, because uplink and downlink data transmission in the TDD system work on the same frequency bands. In TDD, the CSI is obtained only by utilizing the uplink pilot sequence because the downlink and uplink channel matrix are joined. Assuming the Rayleigh fading channel, the communications system archetype can be represented as:

$$y = Hx + N \qquad (6.23)$$

The received vector y is required for evaluating the channel through different channel estimation techniques.

6.7   TRAINING-BASED CHANNEL ESTIMATION

This estimation method sends the trained/pilot signal prior to data transmission at the transmitter end. The pilot signal is the reference signal utilized by the transmitted and received end of the communications system. The receiver utilizes these trained sequences to evaluate the channel parameters and reduce the channel effects. Uplink pilots are orthogonal from one end of the terminal to the other. In each cell, the maximum number of different orthogonal pilot sequences that can be allotted to UT is expressed as:

$$\text{Pilots}_{max} = \text{Coherence interval/Channel delay spread}$$

Errors occur because the noise in the channel is reduced, utilizing the MMSE at the receiver. The transmitted symbol corresponding to the respective UT is multiplied with the channel matrix, and the AWGN signal is added to the resulting channel, thus corresponding to the receiving signal y, which is utilized to evaluate the channel parameters derived from the pilot sequences. The MMSE estimator neutralizes interferences by selecting an appropriate weight w, which can be calculated as:

$$w = \sigma^2 n . \sigma^2 s . I + H\left(H^{-1}\right)H \qquad (6.24)$$

where
   $\sigma^2 n$ is the variance of the noise signal,
   $\sigma^2 s$ is the variance of the transmitted signal, and
   I is the $N_r \times N_r$ size identity matrix.

The estimated transmit symbol ($\hat{x}$) is obtained by multiplying the received symbol with the weight (w), then comparing $\hat{x}$ with the transmitted symbol (x) to estimate the BER. Results can show that the mMIMO system provides superior BER compared to conventional MIMO systems.

CE based on the pilot sequences can be attained by appending pilot symbols either in every OFDM symbol's subcarrier with a definite period or in every single OFDM symbol. OFDM systems require a huge data transmission rate and low BER which can be accomplished through estimators of low complexity and high precision.

Single-dimension (1D) estimators are used to attain agreement between complexity and accuracy. The block-type pilot and the comb estimator are the two types of 1D estimators.

The block-type pilot CE method observes the fast-fading channel. This method is effective when the transfer function of the channel does not alter quickly. This estimator can be derived from either the least-squares method or MMSE.

The comb-type pilot CE method observes the fast-fading channel. In this CE, every single OFDM symbol has several pilot tones. This method is effective in a varied environment. Examples of this type of estimator are the least-squares estimator with a single-dimension interpolator and MLE.

The challenges that the 1D estimator need to address are:

- Pilot contamination
- Low complexity and monitoring capacity.

In practice, the OFDM system produces a bi-dimensional signal, so the estimator evaluates the channel properties in time and frequency domains to decreases average quadratic error. The optimum 2D estimator can be derived from the 2D Wiener filter interpolator. Practically, this estimator increases hardware complexity.

The TBCE decreases the performance of the communications system as the number of antennas is increased, so this estimator is complex in the mMIMO system. Such low-SNR environments call for more training symbols, which would reduce the effective data transmission rate.

6.7.1 Blind Channel Estimation

The blind channel estimation (BCE) technique utilizes the statistics of the incoming data to evaluate the channel parameters without requiring pilot sequences. The BCE method mitigates pilot contamination in a large antenna system. The BCE depends on the natural constraints of the channel for CE and thus cancels the requirement of pilot sequences. This makes the BW economical and enhances the accuracy of the CE. With the Rayleigh fading channel model, the channel parameters are evaluated without utilizing the pre-estimated channel parameters. The BCE algorithm proposed for space-time block coding systems yields better performance in terms of BER versus SNR.

The methodology used by the BCE in the mMIMO system depends on reducing the kurtosis cost function as much as possible. First, this estimator performs mathematical operations such as calculation of orthogonal unitary matrix, determining the eigenvectors of the received symbols to estimate the channel. The transmitted symbols can be reconstructed using the ZF equalization method, which is dependent on the unitary matrix. The BER is then evaluated on the reconstructed data signal.

Accurate CSI is crucial for achieving the benefits of mmWave-mMIMOsystems; however, the task of CE is more difficult in mmWave-mMIMO systems than MIMO systems for the following reasons:

- high pilot overheads
- the requirement of CSI at the transmit and receive end terminals for the precoding and combining in the downlink and uplink, respectively
- huge hardware requirement
- low SNR prior to the beamforming increases the Doppler shift.

In view of these difficulties, the new concept of compressive sensing has been developed.

6.7.2 Compressive Sensing-Based CE Scheme

In practice, the correlation between the signals states that the data rate of these signals should be less than their BW, and also the signal dimension should be greater than the degree of freedom. This shows that these signals possess sparsity in transformation domains. The theory of CS states that the sparse signal can be utilized to reconstruct the message data signal from fewer samples than are required by the Shannon-Nyquist sampling theory. CS-based CE in mmWave-mMIMO systems can be accomplished in three stages. They are:

1. Converting the channel measurement matrices into sparse matrices.
2. Compressing the sparse signals into low dimension signals.
3. Reconstructing the message data signal from the compressed signals.

CS theory determines the large-dimension channel parameters from small dimension measurements through the transformation matrix, which converts the channel measurement signals into sparse signals; thus this method has a low complexity estimator and good system performance.

6.8 CONCLUSIONS

The mmWave-mMIMO technology in the 5G-IoT environment will present network intelligent services with the goal of 'information a finger away, everything in touch'. This chapter has discussed the significant techniques that affect the physical link layer functioning of the mmWave-mMIMO system. The efficient waveform design will open up advanced frontiers in the communications system, supporting high flexibility to enhance the potential of 5G networks. The mmWave-mMIMO system increases the data transmission rates in manifolds through SM. Efficient channel modeling and channel estimation techniques are required for the 5G network intelligent services. The channel estimation in the mmWave-mMIMO system is challenging because of massive antennas, low SNR prior to precoding and the hybrid MIMO transceiver. Precoding reduces the interference and fading effects by focusing the energy beam to the desired user terminal, decreasing the latency on the air interface and improving the spectral efficiency of the network. The hybrid precoding technique is practicable and suitable for the mMIMO system deployed in the mmWave regime because it provides superior system performance with low BER, high antenna array gain and low hardware complexity. However, some technical difficulties need to be solved so that the 5G wireless technology will be capable of supporting hundreds of devices with varying speed, BW and QoS requirements.

ABBREVIATIONS

3GPP	Third generation partnership project
5G	Fifth-generation
ADC	Analog to digital convertor
AOA	Angle of arrival
AOD	Angle of departure
BCE	Blind channel estimation
BER	Bit error rate
BS	Base station
BD	Block diagonalization
BW	Bandwidth
CE	Channel estimation
CMT	Cosine modulated multitone
CP	Cyclic prefix
CSI	Channel state information
CS	Compressive sensing
DAC	Digital to analog convertor
D2D	Device to device
eMBB	Enhanced mobile broadband
FBMC	Filter bank multicarrier
FDD	Frequency division duplex
FMT	Filtered multitone
GFDM	Generalized frequency-division multiplexing
LP	Linear precoding
LOS	Line of sight
MF	Matched filter
MLE	Maximum likelihood estimator

MMSE	Minimum mean squared error
mMIMO	Massive multiple input multiple output
mMTC	Massive machine-type communications
mmWave	Millimeter wave
M2M	Machine to machine
MS	Mobile station
MU	Multiple user
NLOS	Non-line of sight
NLP	Nonlinear precoding
NR	New radio
OFDM	Orthogonal frequency-division multiplexing
OOBE	Out-of-band emission
OQAM	Offset quadrature amplitude modulation
PARP	Peak-to-average power ratio
PPP	Poisson point process
PS	Phase shifter
RAT	Radio access technology
RF	Radio frequency
SCM	Stochastic channel model
SIC	Successive interference cancellation
SMT	Staggered modulated multitone
SM	Spatial multiplexing
SNR	Signal to noise ratio
S/P	Serial-to-parallel convertor
STC	Space-time codes
SVD	Singular value decomposition
SU	Single-user
TDD	Time division duplex
UFMC	Universal-filtered multicarrier
URLLC	Ultra-reliable and low-latency communications
V- BLAST	Vertical bell laboratories layered space-time
UT	User terminal
ZF	Zero-forcing

REFERENCES

1. Lalit Chettri and Rabindranath Bera. 2020. A comprehensive survey on the internet of things (IoT) towards 5G wireless systems. *IEEE Internet of Things Journal*, Vol. 7, no. 1:16–32.
2. D. Tse and P. Viswanath. 2005. *Fundamentals of Wireless Communications*. New York: Cambridge University Press.
3. Kinza Shafique, Bilal A. Khawaja, Farah Sabir, Sameer Qazi and Muhammad Mustaqim. 2010. Internet of things (IoT) for next-generation smart systems: a review of current challenges, future trends, and prospects for emerging 5G-IoT scenarios. *IEEE Access On a Special Section On Antenna and Propagation for 5G and Beyond*, Vol. 8:2302–40.
4. Ehsan Olfat and Mats Bengtsson. 2015. Joint channel and clipping level estimation for OFDM in IoT-based networks. *IEEE Transaction on Signal Processing*, Vol. 65, no. 18:4902–11.
5. A. L. Swindlehurst, E. Ayanoglu, P. Heydasri, and F. Capolino, 2014. Millimeter-wave massive MIMO: the next wireless revolution? *IEEE Communications Magazine*, Vol. 52, no. 9:56–62.
6. 5G Forum, White Paper. 2016. 5G vision, requirements, and enabling technologies. South Korea.
7. Z. Pi and F. Khan. 2011. An introduction to millimeter-wave mobile broadband systems. *IEEE Communications Magazine*, Vol. 49, no. 6:101–7.

8. S. Sun, T. S. Rappaport, R. W. Heath, A. Nix and S. Rangan. 2014. MIMO for millimeter-wave wireless communications: beamforming, spatial multiplexing, or both? *IEEE Communications Magazine*, Vol. 52, no. 12:110–21.

9. T. E. Bogale, X. Wang and L. B. Le. 2017. mmWave Communications enabling techniques for 5G wireless systems: a link-level perspective. *mmWave Massive MIMO: A Paradigm for 5G*, S. Mumtaz, J. Rodriguez, and L. Dai, Eds. 195–225, San Diego, CA: Academic Press.

10. L. Wei, R. Q. Hu, Y. Qian and G. Wu. 2014. Key elements to enable millimeter wave Communications for 5G wireless systems. *IEEE Wireless Communications*, Vol. 21, no. 6:136–143.

11. E. G. Larsson, O. Edfors, F. Tufvesson, and T. L. Marzetta. 2014. Massive MIMO for next generation wireless systems. *IEEE Communicationss Magazine*. Vol. 52, no. 2:186–195.

12. Huawei, HiSilicon. 2016. OFDM based flexible waveform for 5G. 3GPP Standard Contribution. R1-162152. https://www.3gpp.org/DynaReport/TDocExMtg--R1-84b--31661.htm

13. S. Mumtaz, J. Rodriguez, and L. Dai. 2017. *Introduction to mmWave massive MIMO. mmWave massive MIMO: A paradigm for 5G*. London, UK: Academic Press.

14. Qualcomm Inc. 2016. 5G waveform requirements. 3GPP Standard Contribution. R1-162198. https://www.3gpp.org/DynaReport/TDocExMtg--R1-84b--31661.htm

15. X. Zhang, L. Chen, J. Qiu and J. Abdoli. 2016. On the Waveform for 5G. *IEEE Commununication Magazine*, Vol. 54, no. 11:74–80.

16. Huawei, HiSilicon. 2016. 5G waveform: requirements and design principles.3GPP Standard Contribution. R1-162151. https://www.3gpp.org/DynaReport/TDocExMtg--R1-84b--31661.htm

17. Rohde & Schwarz. 2016. 5G waveform candidates. *Technical report*. https://cdn.rohdeschwarz.com/pws/dl_downloads/dl_application/application_notes/1ma271/1MA271_0e_5G_waveform_candidates.pdf

18. T. Hwang, C. Yang, G. Wu, S. Li and G. Y. Li. 2009. OFDM and its wireless applications: a survey. *IEEE Transactions on VehicularTechnology*, Vol. 58, no. 4:1673–94.

19. L. Zheng and D. N. C. Tse. 2003. Diversity and multiplexing: a fundamental tradeoff in multiple-antenna channels. *IEEE Transactions on Information Theory*, Vol. 49, no. 5:1073–96.

20. Gerard J. Foschini. 1996. Layered space-time architecture for wireless communications in a fading environment when using multiple antennas. *Bell Labs Technical Journal*, Vol. 1, no. 2:41–59.

21. A. Alkhateeb, J. Mo, N. Gonzalez-Prelcic, and R. W. Heath. 2014. MIMO precoding and combining solutions for millimeter-wave systems. *IEEE Communications Magazine*, Vol. 52, no. 12:122–31.

22. T. A. Thomas and F. W. Vook. 2014. Method for obtaining full channel state information for RF beamforming. *IEEE Global Communications Conference*. https://doi.org/10.1109/GLOCOM.2014.7037349

23. V. Venkateswaran and A. J. V. Veen. 2010. Analog beamforming in MIMO Communications with phase shift networks and online channel estimation. *IEEE Transactions on Signal Processing*, Vol. 58, no. 8:4131–43.

24. F. Gholam, J. Via and I. Santamaria. 2011. Beamforming design for simplified analog antenna combining architectures. *IEEE Transactions on Vehicular Technology*, Vol. 60, no. 5:2373–78.

25. M. Kim and Y. H. Lee. 2015. MSE-based hybrid RF/baseband processing for millimeter-wave Communications systems in MIMO interference channels. *IEEE Transactions on Vehicular Technology*, Vol. 64, no. 6:2714–20.

26. C. E. Chen. 2015. An iterative hybrid transceiver design algorithm for millimeter wave MIMO systems. *IEEE Wireless Commununicationletter*, Vol. 4, no. 3:1–1.

27. S. Han, I. Chih-Lin, Z. Xu and C. Rowell. 2015. Large-scale antenna systems with hybrid precoding analog and digital beamforming for millimeter-wave 5G. *IEEE Communications Magazine*, Vol. 53, no. 1:186–94.

28. L. Liang, W. Xu and X. Dong. 2014. Low-complexity hybrid precoding in massive multiuser MIMO systems. *IEEE Wireless Communications Letter*, Vol. 3, no. 6:653–56.

29. T. L. Marzetta. 2006. How much training is required for multiuser MIMO? *Proceedings of the 40th Asilomar Conference on Signals, Systems and Computers*, Pacific Grove, CA, pp. 359–363.

30. O. Elijah, C. Y. Leow, T. A. Rahman, et.al. 2015. A comprehensive survey of pilot contamination in massive MIMO-5G system. *IEEE Communications Surveys and Tutorials*, Vol. 18, no. 2:905–23.

31. Y. Liu, Z. Tan, H. Hu, et.al. 2014. Channel estimation for OFDM. *IEEECommunicationss Surveys and Tutorials*, Vol. 16, no. 4:1891–1908.

32. H. Bolcskei, R. W. Heath and A. J. Paulraj. 2002. Blind channel identification and equalization in OFDM-based multiantenna systems. *IEEE Transactions on Signal Processing*, Vol. 50, no. 1:96–109.

33. Y. C. Eldar and G. Kutyniok. 2012. *Compressed Sensing: Theory and Applications*. Cambridge: Cambridge University Press.
34. M. R. Akdeniz, Y. Liu, M. K. Samimi, et.al. 2014. Millimeter-wave channel modeling and cellular capacity evaluation. *IEEE Journal on Selected Areas on Communications*, Vol. 32, no. 6:1164–79.
35. D. Gesbert, H. Bolcskei, D. Gore, et.al. 2002. Outdoor MIMO wireless channels: models and performance prediction. *IEEE Transactions on Communications*, Vol. 50, no. 12:1926–34.
36. A. Alkhateeb, G. Leus and R. W. Heath. 2015. Compressed sensing based multi-user millimeter-wave systems: how many measurements are needed? *Proceedings of the IEEE International Conference on Acoustics, Speech and Signal Processing (ICASSP)*, Brisbane, Australia. http://arxiv.org/abs/1505.00299.
37. Z. Gao, L. Dai, Z. Wang, etnal. 2015. Spatially common sparsity-based adaptive channel estimation and feedback for FDD massive MIMO. *IEEE Transactions on Signal Processing*, Vol. 63:6169–83.
38. F. Bohagen, P. Orten and G. Oien. 2007. Design of optimal high-rank Line-of-sight MIMO channels. *IEEE Transactions On Wireless Communicationss*, Vol. 6, no. 4:1420–25.

7 Algebraically Constructed Short Sequence Families for 5G NOMA Techniques

Lazar Z. Velimirovic
Mathematical Institute SANU, Belgrade, Serbia

Svetislav Maric
University of California San Diego, San Diego, USA

CONTENTS

7.1 INTRODUCTION

According to multiple studies by wireless operators and wireless original equipment manufacturers (OEMs), global mobile data traffic will, by the year 2020, have risen to more than 30 Exabyte (EB) per month, an eightfold increase over 2015, while according to a Cisco White Paper [1], total worldwide data traffic has grown 18 times since 2016. The estimated combined number of mobile devices and connections in 2016 reached 8 billion, with mobile networks carrying 7.2 EB per month, while the predicted exponential boom of connected things (e.g., sensors and wearable devices) is expected to reach over 75 billion by 2025 [2]. Furthermore, a plethora of contemporary applications and use cases are expected to be generated with the development of a new generation of wireless communication networks, like massive machine-type communications (e.g., e-health, millions of sensors connected, smart home), enhanced mobile broadband (e.g., virtual reality, augmented reality) and ultra-reliable and low latency communications (e.g., remote surgery, vehicle-to-everything communication, self-driving cars, drone delivery, smart manufacturing) [3].

The fifth-generation wireless system (5G), officially the International Mobile Telecommunications-2020 (IMT-2020 Standard for 5G Networks, Devices and Services), demonstrates the next important chapter in mobile telecommunications standards beyond 4G. IMT-Advanced

DOI: 10.1201/9781003045809-9

Standards advance present mobile broadband services, but additionally include the expansion of mobile networks to enable an ever-growing diversity of services and applications as well as the number of connected devices. It is evident that the coming 5G network will have to deal with high traffic volumes and computing demands.

In order to address these challenges, there are some ever-present expectations from the 5G cellular systems, including (1) 1000-fold higher system capacity per sq.km., (2) up to 20GB/s transmission rate, (3) connectivity of almost 100 billion devices, (4) zero latency, less than 1ms over the RAN, (5) maximum mobility speed of almost 500km/h, (6) 99.999% reliability and maximum coverage, and (6) increase of energy efficiency up to 100 times [4].

The expected massive reception and computing of data in 5G networks will demand efficient processing alongside an enormous increase in energy consumption by the wireless communication infrastructure. To satisfy these needs, a corresponding growth in base stations (BS) is expected [5, 6]. The BS, as a major energy consumer, is estimated to use around 60% of the entire network energy consumption, which is more than 70% of the total electricity bill. Mobile communication, therefore, has a significant share of the total energy consumed by the information technology industry, but also in globally emitted CO_2, making its share above 5%, and it is expected to rise extremely fast. Cellular networks account for CO_2 emissions estimated at between 0.5% and 1% of the global carbon footprint. Besides the ecological and economic issues, the CO_2 levels are contributing to an increase in transmission power levels within cellular networks, causing a rise in radiation levels which affects human health [5, 6].

Operators are seeking innovative approaches to cut down energy use in 5G networks without compromising network coverage and capacity. Nevertheless, many 'greener' strategies, including energy harvesting technologies and renewable energy sources, which should take into consideration environmental, economic and social concerns, are typically either insufficient or act against the constantly growing traffic requirements. Also, the propagation of signals at millimeter-wave range requires a line-of-sight transmission path, and it is highly vulnerable to disruption from physical infrastructure, vehicles and surroundings [5, 6].

5G technology will be using a small-cell concept for transmission, able to deliver high data rates by serving a small number of users under each cell. Consequently, enormous investments are expected for deployment of a massive number of microcells to cover a vast geographical area (e.g., region/country), which calls for an increasing number of BSs. The trade-off between energy efficiency and data traffic will be one of the key challenges for design excellence of the 5G network, and is thus a timely research topic [5, 6].

These types of 5G scenarios are often referred to as machine-type communication (MTC) and can be further divided into two main types: massive MTC with low data rates; and MTC with low latency and high reliability [3]. In this classification, smart city, infrastructures and objects (e.g., massive devices like sensors and actuators) can be classified as massive MTC [7–9], while autonomous vehicle control, smart grid, factor-cell automation and similar services come under MTC with low latency and high reliability [10–12].

For massive MTC, the network is expected to accommodate a large number of connections with sparse short messages [3], which should be low cost and energy efficient to enable large-scale deployment. It is this type of MTC that we discuss in this chapter. We analyze the possibility of using non-orthogonal multiple access (NOMA) techniques for serving MTC traffic in the form of IoT. Our particular interest in MTC comes from the fact that it is a perfect example of how in engineering, theoretical mathematical concepts, even if established many years ago, can still be useful in modern systems (i.e., 5G networks). In fact, in this chapter we will use algebraic sequence construction theorems and different methods for multiple access to illustrate their applications in NOMA.

The chapter is organized as follows. In the next section, non-orthogonal multiple access (NOMA) system, sparse code multiple access (SCMA), pattern division multiple access (PDMA), and multiple-user shared access (MUSA) are explained in detail. Then follows a section in which new ways of constructing sequences and frequency-hopping (FH) codes that can be applied in PDMA and

MUSA systems are proposed. The results obtained are presented, and comments on the achieved performance improvements of the observed systems are followed by conclusions and a review of future challenges on the topic of 5G and IoT.

7.2   NON-ORTHOGONAL MULTIPLE ACCESS (NOMA) SYSTEM

From the 1980s and the introduction of 1G standards to the present day and the development of 4G and 5G standards, the development of several key multiple-access technologies responsible for the reliable operation of wireless communication systems can be observed. Depending on whether we are looking at time, frequency, code domains or some combination of them, or depending on which domain wireless resources are assigned to multiple users, we can distinguish frequency division multiple access (FDMA), time division multiple access (TDMA), code division multiple access (CDMA) and orthogonal frequency division multiple access (OFDMA) [13, 14]. All four multiple-access schemes can be classified into a set of technologies belonging to orthogonal multiple access (OMA).

One of the biggest problems with orthogonal multiple access (OMA) is the limited number of orthogonal resources which automatically leads to a limit on the number of users that can be supported. Another problem is reflected in the loss of orthogonality due to channel-induced impairments, regardless of the corresponding orthogonal resources.

To overcome these two main shortcomings, a new concept called the non-orthogonal multiple access (NOMA) system has been proposed [15, 16]. When designing the NOMA system, the main idea was to provide support to a larger number of users compared to the number of orthogonal resources, at the expense of increasing the complexity of the receiver. Power-domain NOMA and code-domain NOMA represent two categories of solutions that can achieve the set goals [17–20].

Several multiple-access schemes based on the principles of NOMA technologies have been developed over time. First, interleave-division multiple access (IDMA) has been proposed, which is based on CDMA technology [21]. Subsequently, low-density spreading code division multiple access (LDS-CDMA) was developed, which was enhanced by the development of low-density spreading-aided orthogonal frequency division multiplexing (LDS OFDM) [22, 23]. LDS OFDM actually represents the synthesis of LDS-CDMA and OFDM. Further modification of the LDS OFDM developed two more multiple-access schemes: sparse code multiple access (SCMA) and successive interference cancellation amenable multiple access (SAMA) [24–26]. Pattern division multiple access (PDMA) was developed by improving SAMA [27, 28]. Bit-division multiplexing (BDM) and multi-user shared access (MUSA) have also been proposed as solutions based on the principles of NOMA technologies [29, 30]. In BDM theory, hierarchical modulation is respected, whereby resources are divided not to symbol level but to bit level. MUSA is an advanced CDMA scheme which means it is based on code-domain multiplexing.

NOMA is just one of the potential candidates in the physical layer, which can increase system throughput and simultaneously serve massive connections as required in MTC. Of course, in order to satisfy all the 5G use cases as defined by 3GPP, many more enhanced or revolutionary technologies are currently being developed. For instance, [31, 32] discuss the use of C-RAN to efficiently bring user information closer to the main network processor, increasing system capacity and at the same time increasing user equipment speed, since it is not necessary for it to perform complex operations.

In general, NOMA techniques can be classified into two different types. The first type is represented by power and rate splitting [16–19]. These techniques are known to achieve sum rate capacity of the system, and introduce non-orthogonality via power-domain user multiplexing. Thus, power domain is the basis of how NOMA is defined. On the transmitter side, NOMA users are multiplexed in the power domain, while from the receiver side, NOMA users are separated by SIC [33].

This NOMA system can be used for downlink and uplink and has been researched for increasing long-term evolution (LTE) capacity.

The second NOMA category, used mainly for uplink in the context of MTC and general IoT, is characterized by new multiple-access techniques. As described below, these techniques fall into the category of non-orthogonal multiple access because the addresses used to distinguish the users are non-orthogonal. We recall that in 3G CDMA systems, each user through spreading utilized the whole-time frequency (uplink or downlink) space. The spreading sequences used as users' addresses were orthogonal, to minimize user interference. Since the users utilize the large system time frequency space, there is practically no limit to the length of the individual user codes, ensuring ideal address (code) orthogonality.

For this second NOMA category, the underlying system is still the LTE time frequency space, but multiple access is achieved on the resource element (RE) level [20]. As in the CDMA, user interference is minimized by the use of nearly orthogonal patterns and/or sequences. The reason the addresses are nearly orthogonal (quasi-orthogonal), is that the RE has very limited time duration and narrow bandwidth. This constraint limits the length of the addresses, as well the corresponding family size, necessitating construction of best possible quasi-orthogonal sequences.

Since NOMA, unlike CDMA, was created as an 'overload' system, it is not possible to create completely orthogonal codes because the number of REs, in a system like this, is fewer than the number of users [20].

One general example (meaning no specific access technique) is illustrated in Figure 7.1 [3]. A new class of protocols, based on coded random access, provides massive coordinated access [34].

In Figure 7.1 there are collisions in slot 1 between devices 2 and 3, and in slot 3 between devices 1 and 3. Using classical detectors (with or without error-correction coding), most of the information in those slots would be lost (and in the best case marked for retransmission). Using a receiver based on successive interference cancellation (SIC), it is possible to decode the colliding packets. Some SIC decoding techniques are very complex and require extensive processing. However, basic SIC principles can be understood using Figure 7.1. If the receiver applies SIC, it can store the collided packets from slots 1 and 3. The decoded packet from device 2 (that had no collisions) in slot 4 contains a pointer to where else device 2 has transmitted. The receiver then cancels the interference of device 2 from the buffered reception from slot 1 and thus recovers the packet from device 3. Finally, it cancels the packet from device 3 from the buffered reception in slot 3, thereby recovering the packet of device 1 [3]. The above technique also illustrates the fact that complex processing is required in this case, first for storing and memorizing the collided packets, and then for cancelling the power of the interferer.

Successful decoding of transmitted symbols depends on eliminating the error propagation in the receiver. Error propagation occurs due to the inability of the receiver to decode the symbol based on decoding of only one received RE (this is clearly so because of simultaneous use of the RE by more

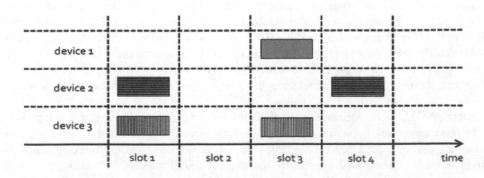

FIGURE 7.1 Typical coded random access.

than one user). The receiver, therefore, needs to wait for the information in the next RE and try to perform decoding.

This leads to the need to design advanced complex detection algorithms such as the SIC (discussed above) or a belief propagation (BP) receiver. Design complexity can render practical applications of NOMA prohibitively expensive. One way to reduce receiver complexity is to design relatively sparse user patterns that can be decoded using BP algorithms. These algorithms have demonstrated encouraging performance for overloaded systems like NOMA.

The most promising NOMA candidates for MTC are: sparse code multiple access (SCMA), pattern division multiple access (PDMA) and multi-user shared access (MUSA) [35, 36].

7.2.1 SPARSE CODE MULTIPLE ACCESS (SCMA)

Sparse code multiple access (SCMA) is a NOMA technique realized on sparse structure codebook principles [24]. In 5G, this technique can be applied both in downlink and uplink transmission.

Selecting the entries of a signature sequence matrix S for optimal efficiency for use in BP is addressed in [37]. The traditional maximum a posteriori (MAP) receiver bases its work on the MAP detector:

$$\widehat{x}_{MAP} = \arg\max_{x \in X^K} \prod_{n=1}^{N} M_n(x),$$ (7.1)

where:

$$M_n(x) = \exp\left\{-\frac{1}{\sigma_\omega^2}\left|y(n) - \sum_{k=1}^{K} x_k s_k(n)\right|^2\right\},$$ (7.2)

is the chip-metrics. In order for the sequences to be optimal for the SCMA systems, first the non-zero values in S must only occur in the BP algorithm through $M_n(x)$. In order for the sum-product message passing to perform correctly in the additive white Gaussian noise channel, the metric should be characterized by discriminating features with regard to the received symbol alternatives in the received N-dimensional constellation $\{Sx \mid x \in X^K\}$. The first condition that has to be fulfilled is to ensure unique decodability, while the second condition is to ensure that the received vectors from the set $\{Sx \mid x \in X^K\}$ are at the greatest possible minimum Euclidian distance between them. In [38] the use of a low-density signature matrix for codes in SCMA is investigated. The matrices, when developed in the context of LDPC codes, have the sparse property, namely the number of 1s in an otherwise null matrix is very small. The position of the non-zero entries is indicated by a low-density spreading (LDS) indicator matrix. The applicability of the codes comes from the similar requirements for decoding in LDPC and SCMA and depends on the sparsity of the matrix.

It is important to note that an indicator matrix only indicates the position of non-zero elements, (denoted by '1'), but not their values. If an indicator matrix is to be used as a signature sequence in a multiple-access system, then not only a proper mapping of '1' elements to different users has to be achieved, but also proper value assignment of the corresponding signal.

In [38] the problem of assigning '1s' to users by the use of the Latin square property is solved [39]. As we will see in the section on PDMA, that property is shown to be optimal also in our design of PDMA patterns.

From the implementation perspective, in order to assign values to the chips it should possess constant-modulus chip constellations, as well as all values in each row being distinct.

With that in mind, the chip values are assigned as follows: a_k, $k = 0, 2, \omega_r - 1$, of the signature constellation A as

$$a_k = \exp\left(j \frac{2\pi}{P} k \right), \quad k = 0, 1, \ldots, \omega_r - 1, \quad P > \omega_r \tag{7.3}$$

Here:

$$P = q\omega_r \frac{\min(q, \omega_r)}{\gcd(q, \omega_r)}, \tag{7.4}$$

where $q = |X|$ is the size of the symbol constellation X and $\gcd(q, \omega_r)$ is the largest integer dividing both q and ω_r without remainder, producing a uniquely decodable system for all of the large number of LDS indicator matrices of various sizes and various information constellations X.

7.2.2 PATTERN DIVISION MULTIPLE ACCESS (PDMA)

A novel multiple-access scheme based on the principles of NOMA technologies is PDMA [28]. In order to be able to perform multiplexing in one of the three domains (code, power, space) or in some possible combination of these domains, PDMA uses non-orthogonal patterns on the transmitter side. Non-orthogonal patterns meet the following conditions: diversity is at the maximum level, while the overlaps of multiple users are the minimum possible.

Viewed from the code-domain angle, the difference between multiplexing in PDMA and In SCMA is that the same symbols can be associated with a different number of subcarriers. To eliminate interference on the receiver side, MPA is applied. Viewed from the angle of the power domain, the total power constraint is the main condition that should be taken into account during power allocation. Spatial PDMA, or observing from the angle of the space domain, is very useful because it allows successful combination with the multi-antenna technique. Compared to MU-MIMO, the application of spatial PDMA significantly reduces system complexity because spatial orthogonality can be achieved without joint precoding. Also, to maximize the various wireless resources available, it is possible to combine multiple domains in PDMA. In order to decide on the greatest number of supported users with N orthogonal subcarriers of one uplink PDMA system based on code domain, we will consider an overloaded system, and we will assume that the number of users K is higher than the number N which represents the orthogonal subcarriers of OFDM [28].

In code-domain PDMA, the non-zero elements of any spreading sequence b_k for user k are equal to 1. The spreading matrix marked with $B = (b_1, b_2, \ldots, b_{K_2})$ is designed according to the following conditions:

- The number of groups that in the spreading sequence contain a different number of 1s should have the highest value.
- The number of groups that in the overlapped spreading sequences include the same number of 1s should have the lowest value.

When these conditions are fulfilled the maximum number of supported users with N orthogonal subcarriers equals:

$$\binom{N}{1} + \binom{N}{2} + \ldots + \binom{N}{N} = 2^N - 1. \tag{7.5}$$

As discussed above in the PDMA system, users are divided by short non-orthogonal patterns on the transmitter side. It relies on the notion of 'code sparsity' to reduce the complexity of the receiver and hence worth implementing a BPA receiver. For example, code or spatial domain represent domains in which it is possible to implement the PDMA technique, as long as the patterns can be separated in the respective domain.

Since the number of REs is less than the number of users, PDMA belongs to the overloaded system. Combining Lagrange or Sidelnikov sequences with polynomials over finite fields, we create patterns and obtain sequences based on the quasi-orthogonal nature. Sequence properties, i.e., dependency of BER on SNR for PDMA, are determined using link-level simulations.

In Equation (7.6) we give an example of a general form pattern matrix with $m = 3$ users in $n = 2$ REs. Here each column is one RE, and each raw represents the user RE utilization.

$$U = \left(u_{i,j}\right)_{m \times n} , \; i = 1,\ldots,m; \; j = 1,\ldots,n. \tag{7.6}$$

Equation (7.7) gives a specific example of (7.6) and is a 150% overloaded PDMA system.

$$U = \begin{pmatrix} 1 & 0 \\ 1 & 1 \\ 0 & 1 \end{pmatrix}. \tag{7.7}$$

As shown in Equation (7.7), patterns of three users are described with three rows. The first RE gives information about the first user, the second RE gives information on the third user, whereas information on the second and the third users is contained in both the first and second RE. At the receiver, SIC separates different users' signals.

7.2.3 MULTIPLE-USER SHARED ACCESS (MUSA)

In addition to low energy consumption and acceptable cost of devices, for the successful implementation of IoT it is necessary to provide low latency and a huge number of simultaneous connections. Another type of NOMA scheme that can respond to these IoT requirements is multi-user shared access (MUSA) [30]. One of the principles on which MUSA is based is grant-free access. Thanks to the grant-free access procedure, the access procedure is greatly simplified. Also, a new way of code-domain spreading allows a huge number of users to be accepted in the same radio resources. Also, due to the grant-free access and non-orthogonal multiple-access scheme for data transmission, MUSA successfully solves the problem that at the same time the minimum energy consumption and signaling overhead on the one hand and the maximum number of connections on the other are achieved. Without grant-free access, i.e., with grant-based transmission, the spectral efficiency of transmissions would be very poor, while transmission delay and signaling overhead would be huge if multiple users required simultaneous access to the network. The simplest way to implement grant-free access is to use random spreading codes. In order to achieve better system performance, a special family of complex sequences is created.

The data that characterize each user are spread using a family of complex sequences with short length. At the receiver, signals from many users are superimposed, using SIC for interference cancellation between users. Simple SIC is enabled by a specific construction of the spreading sequence, which also successfully overcomes the problem reflected in the high user load [30].

For uplink MUSA the same spreading sequence multiplies all the transmitted symbols of one user. Also, in order to reduce interference, for the same user, different spreading sequences can be applied to different symbols. After that, the same time frequency resources are used to transmit all symbols.

In MUSA the transmitter forms a codebook (CB) that contains, as elements, R different symbols in a complex plane: $CB = \{s_1, s_2, \ldots, s_r\}$. Each user, i, is assigned a codeword, cw_i of predefined length m, $m \leq R$, consisting of m different complex symbols from the CB. This codeword then modifies a MUSA complex carrier, and acts as a non-orthogonal spreading code. As mentioned above, a MUSA system is also an overloaded system, which means that the codeword length is less than the number of users.

7.3 SEQUENCE CONSTRUCTION: MODIFY FREQUENCY HOP CODES AND LAGRANGE SEQUENCES FOR MUSA AND PDMA SYSTEMS

Patterns for effective PDMA location require small inner items that lead to more modest interference [40]. The sequences we propose are for the short lengths—almost orthogonal, and thus appropriate for PDMA use.

On the other hand, in MUSA the sequences act as a short complex spreading code. Therefore, interference in the system is created by non-ideal cross-correlations between sequences. Interference can be minimized using successive interference cancellation in the receiver; however, this adds to the receiver complexity. It is much more efficient to design the codes to be as orthogonal as possible and in turn reduce receiver complexity [41].

Our development comprises two sections, and combines two ideas for building sequences. The initial segment develops PN sequences (for example, Legendre, Sidelnikov [42, 43]), that have perfect autocorrelation characteristics [44]. The characteristic of these sequences is that they are based on prime numbers. For each prime a low number of the listed sequences can be found. In fact, because of Lagrange sequences only one sequence for each prime can be found. The latter segment of our development raises the number of sequences in the sense that the original sequence shifts cyclically in a very specific way using permutations designed for frequency hopping (FH) in communications systems with many users [45]. Almost perfect two-dimensional correlation functions belonging to FH codes [45] when applied as shift sequences to Lagrange or Sidelnikov sequences will guarantee that the recently designed sequences' family possesses a quasi-orthogonal nature.

It is important to note that the construction is essentially based on original sequence shifting by quasi-orthogonal permutations.

To create two different users we will use these two sequences as patterns. The length of the FH permutation directly influences the number of patterns, while prime directly influences the length of the FH permutation. Generalized Welch sequences will be used for our shifting permutation [43].

7.3.1 SIDELNIKOV, LEGENDRE, AND COMPLEX LEGENDRE SEQUENCE DEFINITION

In wireless communications systems, sequences with low correlation are widely applied. This feature of the sequence is very important because it can be used, for example, in cases of low mutual interference between multiple users or channels or in a situation of obtaining correct timing information. It is important to note here that not all wireless and communication systems in general impose similar conditions for sequence properties.

Systems can have different requirements and some are listed below:

- Good one-dimensional synchronous and asynchronous acyclic autocorrelation and cross-correlation.
- Good two-dimensional synchronous and asynchronous acyclic autocorrelation and cross-correlation.
- Good one- and two-dimensional synchronous and asynchronous cyclic autocorrelation and cross-correlation.

In the case of adaptive modulation schemes, the important role has variable alphabet sizes of the considered sequences. The application of such sequences enables the harmonization of channel characteristics and variable data rates, which leads to improved performance of the entire wireless system.

One type of sequence that meets these conditions is the Sidelnikov sequence [46]. Sidelnikov sequences have variable alphabet sizes and low correlation. They are polyphase sequences characterized by multiplicative characters. In [47], the binary nature of these sequences is shown.

When shift and addition is performed over Legendre sequences of prime period p, binary sequences with important correlation properties are obtained [48]. These features allow, using Sidelnikov sequences, creation of a large number of different polyphase sequence families.

Having in mind that the value of the correlation magnitude of the sequences is $2\sqrt{p} + 5$, M-ary sequence families based on Sidelnikov sequences are created, over which shift and addition was applied [49, 50].

For the same value of maximum correlation magnitude, using the same shift-and-addition procedure, [51] explains the creation of new M-ary sequence families that are significantly larger than the previously created M-ary sequence families.

Using the refined Weil bound, in [52] the general procedures for constructing the already known M-ary sequence families are presented, giving new evidence of the value of the maximum correlation magnitudes. Also, the procedure in which Sidelnikov sequences perform multiple cyclic shifts served to generalize the constructions of the M-ary sequence families.

The Sidelnikov sequence, under the condition that p represents an odd prime, while α is its primitive root, is defined in the following way:

$$s(t) = \begin{cases} 1, & if \ \left(\alpha^t + 1\right) \in N, \ t = 0,1,\ldots,p^m - 2 \\ 0, & \text{otherwise} \end{cases}. \tag{7.8}$$

In the previous expression, a set of quadratic nonresidues over the finite field $F\{p^m\}$ is marked with N and is equal to:

$$N = \alpha^{2n+1}, \ n = 0,1,2,\ldots,\left(\left(\frac{p^m - 1}{2}\right) - 1\right), \quad m - \text{integer}. \tag{7.9}$$

Legendre sequences and quadratic residue codes have a number of common features that are reflected in randomness properties. The pattern distribution represents one randomness aspect of sequences. The distribution of some special patterns is one of the postulates that Golomb developed [53].

The optimal balance between values 1s and 0s is one of the features that Legendre sequences possess. If $N \equiv 3(\text{mod}4)$, their periodic autocorrelation function is two-valued, whereas for the case where $N \equiv 1(\text{mod}4)$ this function is three-valued, based on which it can be concluded that their autocorrelation property is optimal. In [54], it is shown that, depending on what value N mod 8 has, one of the values from the set $\left\{\frac{N-1}{2}, \frac{N+1}{2}, N-1, N\right\}$ shows the linear span of the Legendre sequence.

The Legendre symbol is defined in the following way:

$$\left(\frac{a}{p}\right) = \begin{cases} 1, & if \ a \text{ is a quadratic residuo mod } p, \ a \equiv 0 \text{ mod } p \\ -1, & if \ a \text{ is a quadratic non residuo mod } p, \\ 0, & if \ a \equiv 0 \text{ mod } p \end{cases}, \tag{7.10}$$

where the operations are performed on $\text{mod}(p)$, and p is an odd prime.

TABLE 7.1

Coefficients s_k of 4 Complex Legendre Sequences of Length 4

		k			
		1	2	3	4
r	1	2	4	3	1
	2	4	3	1	2
	3	1	2	4	3
	4	3	1	2	4

The Legendre sequence will in fact represent the codebook from which the MUSA signals are formed.

(Complex Legendre [43]). A Complex Legendre Sequence consists of the following symbols:

$$s_k = e^{-2\pi j \frac{c_k}{p-1}}, \quad c_k = r * ind_\alpha k, \tag{7.11}$$

where $r \in GF(p) - \{0\}$ and $k = 1, ..., p-1$, (p - odd prime, α its primitive root). The index function 'ind' is equivalent to the log function and the base is the primitive root.

For example, for $p = 5$, $\alpha = 2$, we have $2^4 \equiv 1 (mod 5)$ hence $ind_2(1) = 4$.

For $p = 5$ and $\alpha = 2$, in Table 7.1 we give the values of s_k for different values of r, and k.

7.3.2 GENERALIZED WELCH (GW) SHIFTING SEQUENCE CONSTRUCTION

Today there are two general methods that are based on the theory of finite fields which serve to generate Costas arrays. These are the Welch and the Golomb-Lempel methods [55, 56]. Costas arrays, created by Prof. Costas [57, 58], and whose construction methods were developed by Prof. Golomb, represent frequency-hopping patterns which own the feature of ideal 2D autocorrelations.

This ideal two-dimensional acyclic autocorrelation is one of two properties defining Costas arrays (the second one is that each time frequency resource element has to be utilized once and only once). On the other hand, Costas arrays possess rather poor two-dimensional acyclic cross-correlation properties [59]. Due to these poor cross-correlation properties, they are not suitable for use in multiplexing and multi-user systems because they do not reduce cross-talk enough. In order to solve this problem, special subfamilies with low cross-correlation were developed [60].

For both methods of constructing Costas arrays, the Welch and the Golomb-Lempel methods, there are more submethods based on which desired system performances can be achieved [55, 56].

Generalized Welch (GW) sequence $s(i)$ is described with the following expression [42]:

$$s(i) = \log_a \left(A\alpha^{2i} + B\alpha^i + C \right), \ (\text{mod } p), \ i = 1, 2, ..., p-2, \tag{7.12}$$

where α- primitive element of $GF(p)$, A, B, $C \in GF\{p\}$.

Equation (7.12) is a generalization of the Welch Construction for Costas arrays as used in frequency hop coding [45].

The sequence:

$$s(i) = \left(A\alpha^{2i} + B\alpha^i + C \right), \ (\text{mod } p), \ i = 1, 2, ..., p-2, \tag{7.13}$$

has the length $p - 2$, where the values of elements belong to the set $\{0$ to $p - 1\}$.

TABLE 7.2
Shift Sequence Obtained Using Equation (7.13)

	1	2	3	4	5
0					
1					
2		X			
3	X				
4				X	
5					X
6			X		

For the parameter values $p = 7$, $\alpha = 3$, $B = 1$, and $A = C = 0$, the sequence $s(i)$ equals to $s(i) =$ 3,2,6,4,5, (see Table 7.2).

It is important to note that the value of the autocorrelation of the family of sequences defined by the previous expression is at most two, thanks to their quadratic nature and Lagrange's theorem [7].

7.3.3 CONSTRUCTION OF SHORT PATTERNS FOR PDMA AND MUSA

The method of creating short patterns for PDMA in this chapter will be explained using an example. Let us have a look at the Sidelnikov sequence $g(t)=0,0,1,0,1,1$ and the GW array $s(i) = 3,2,6,4,5$. From the GW array it can be noted that '3' is the first element, which means that the pattern of the first user is obtained by shifting the Sidelnikov sequence for that value. In this case, it equals to $s_1 = 0\ 1\ 1\ 0\ 0\ 1$. Shifting the original Sidelnikov sequence for the value of the second element of the GW array, we obtain the pattern of the second user, and so on (see Figure 7.2). The length of the Sidelnikov sequence determines the weight of the sequence, while the length of the GW array determines the number of users.

In MUSA, we need to form a codebook but since the sequences are complex, we use complex Lagrange sequences as defined in (7.11). If $p = 5$, then the values of c_1 from the first row of Table 7.1 and cyclically shifting them using the first element of the shift sequence from Table 7.2, which is 3, we obtain symbols for the complex signal of the first user:

$$s_1 = e^{-2\pi j\frac{4}{6}},\ s_2 = e^{-2\pi j\frac{3}{6}},\ s_3 = e^{-2\pi j\frac{1}{6}},\ s_4 = e^{-2\pi j\frac{2}{6}}. \tag{7.14}$$

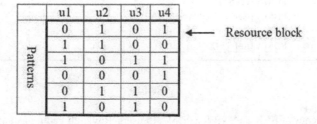

FIGURE 7.2 The example of creating pattern for the case of 4 users.

Similarly, the second user's signal will be obtained by shifting the second row of Table 7.1 with the second element in Table 7.2, and so on. Each of the signals is complex and has 4 different symbols.

It is possible to increase the number of symbols by 1. Note that if we allow $k = 0$, in Equation (7.11), then $s_0 = -\infty$. In [43] it was shown that if $-\infty$ is mapped to symbol '0' the correlation properties of complex Legendre sequences do not change. In this case, the number of symbols is equal to p.

7.4 PERFORMANCE COMPARISON OF DIFFERENT PDMA PATTERNS

The overall PDMA efficiency regarding capacity is assessed in [40]. These performances are determined as dependency of values of the uplink and downlink spectral efficiency parameters on the SNR. Observing for the case one RE, we determine dependency of BER on SNR for our recently built patterns. We likewise contrast their performance with PN pattern examples and examples described in [40].

Figure 7.3 shows the evaluation of pattern capabilities on one RE first with five users, where the overload is 125% (four-length pattern code). Please note that our link-level simulations are carried out for one RE. The user increase due to NOMA is achieved when all the REs are simultaneously in use. Also, this figure shows performances of PN pattern and two patterns described in [40]. Overload of patterns described in [40] are 25% higher with respect to our and PN pattern.

These patterns are taken for comparison, regardless of the slightly higher overload, because it is the only relevant published sequence we identified.

As expected, for the same overload, the patterns have much better performances than those achieved using PN patterns, as well as compared to the patterns used earlier. There are two main reasons for these results, namely lower overload and raised separation properties of patterns we have created.

Pattern performances for the case of ten users where the overload is 250% (four-length pattern code) are shown in Figure 7.4. Based on the results shown, it can be observed that slightly better BER performance on the observed SNR range is achieved using the proposed deterministic sequences compared to PN sequences. As the possibility of maintaining separation between the patterns decreases with the increase of the overload factor, the results shown in Figure 7.4 are expected.

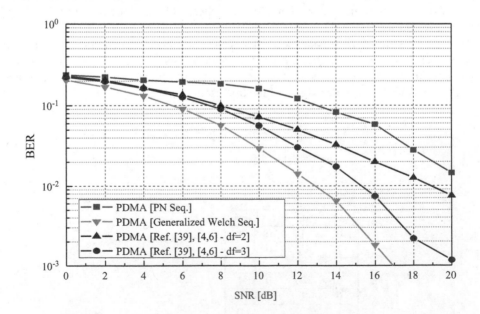

FIGURE 7.3 Dependency of BER on SNR for PDMA (5 users, Rayleigh Fading Channel).

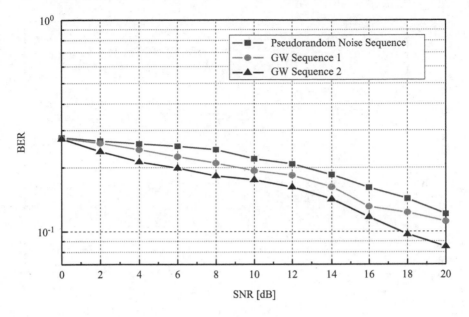

FIGURE 7.4   Dependency of BER on SNR for PDMA (10 users, Rayleigh Fading Channel).

The main advantage of the proposed method of sequence construction is reflected in the high flexibility, both in length and in number. Therefore, in different resource blocks it is possible to use different sequences. The performances shown are achieved for cases where five or ten users are placed in those resource blocks, which are very successfully separated by means of IC. Also, in accordance with the requirements of the system, it is possible to increase the number of required sequences by increasing the prime, in cases when in one resource block a larger number of users are being scheduled.

As with the PDMA the performance is done over one RE. Unfortunately for MUSA we have not been able to find any code construction in the literature, even if [4] states the investigation in the codebook is very active. We therefore compared the constructed sequences with random sequences (see Figure 7.5 and Figure 7.6) and as expected we can see the superior performance of the constructed sequences.

What is more interesting is that if we compare the performance of MUSA and PDMA (Figures 7.3 and 7.4, vs Figures 7.5 and 7.6) the schemes show similar performance for the same overloads and similar SNRs. This is the result of higher diversity achieved by complex spreading codes and is in line with the results in reference [35].

At the end of the chapter it is worth mentioning a technique that enables transformation of one-dimensional sequences to two-dimensional arrays, and vice versa. The technique referred to as folding and unfolding of arrays is based on the Chinese Remainder Theorem, and was introduced by Luke [61] in the context of constructing perfect arrays, and by Moreno and Maric in [62], in the context of constructing frequency hop codes. Other applications of folding sequences are given in [63].

The folding technique for a sequence of length N into an array of size $N_x N_y$ (with $N = N_x N_y$, and N_x, N_y relatively prime) is illustrated in Figure 7.7.

As can be seen from Figure 7.7, the considered sequence consisting of 21 symbols can be converted into a sequence having 3 columns and 7 rows. Observing diagonally the symbols of the displayed sequence that are repeated, the initial periodic repeated sequence can be noticed, which is why it is possible to perform the transformation mentioned. Array unfolding can be used, for instance, in transforming PDMA patterns (when treated as a 2D pattern into sequences for MUSA).

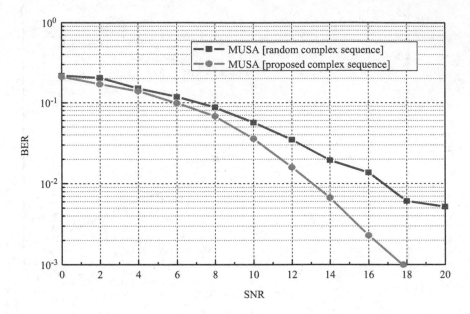

FIGURE 7.5   Dependency of BER on SNR for MUSA (5 users, 4 RE, overload 125%, Rayleigh Fading Channel).

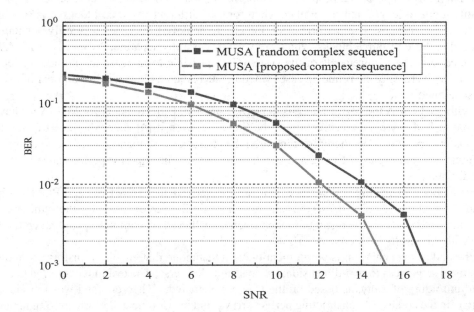

FIGURE 7.6   Dependency of BER on SNR for MUSA (10 users, 4 RE, overload 250%, Rayleigh Fading Channel).

7.5   CONCLUSION

NOMA technology, due to the working principles on which it is based, is a very serious candidate for 5G applications. However, in order to meet and solve many of the accompanying problems for the successful implementation of NOMA for MTC in 5G, it is necessary to modify existing or develop new solutions. Some of the problems to be solved are: design of spreading sequences or codebooks, receiver design, channel estimation, grant-free NOMA, resource allocation, extension to

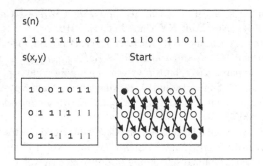

FIGURE 7.7   The folding technique for a sequence of length N into an array of size $N_x N_y$.

MIMO, cognitive radio-inspired NOMA, the reduction of the PAPR and base-station cooperation. Implementing software-defined multiple access technology in NOMA can significantly improve system perforations and expand scope and support in 5G.

IoT MTC in 5G encompasses many diverse use cases, from low latency/high reliability to low rate/uplink sensor traffic. Therefore, it is expected that there will be multiple physical layer techniques, each optimized for a specific set of use cases.

That is also true for the NOMA techniques discussed above. For instance, even though the diversity gain of PDMA and MUSA schemes is relatively low compared to the gain in SCMA, these NOMA schemes still provide good BER performance with low and moderate overloading factors and can be used in uplink IoT. Their performance can be further improved with a careful selection of patterns and sequences as shown above. However, the deciding factor for the applicability of NOMA in IoT MTC will be the availability of a large number of nearly orthogonal sequences. The constructions outlined above solve that problem thanks to the large number of independent coefficients in the generating equations.

SCMA, on the other hand, offers an increased number of users both in the downlink and the uplink and can therefore be used also for data transmission if the reliability is improved.

Another important conclusion from this chapter is that as with previous multiple-access systems, proper sequence selection remains a very important part of a system design. We have also shown that it is better to rely on families of algebraically designed sequences, as individual sequences possess similar characteristics, hence guaranteeing predictable performance.

Simulations also show that by using the algebraically constructed sequences much better performances are achieved with respect to BER on the observed SNR range than with using PN or other sequences. In future work we will evaluate the performance of the sequences on the ITU-defined propagation channels.

ACKNOWLEDGMENT

This work was supported by the Serbian Ministry of Education, Science and Technological Development through the Mathematical Institute of the Serbian Academy of Sciences and Arts.

REFERENCES

1. Cisco, Cisco annual internet report (2018–2023) white paper, 2020, [online]. https://www.cisco.com/c/en/us/solutions/collateral/executive-perspectives/annual-internet-report/white-paper-c11-741490.html.
2. A. Osseiran, F. Boccardi, V. Braun, K. Kusume, P. Marsch, M. Maternia, O. Queseth, M. Schellmann, H. Schotten, H. Taoka, H. Tullberg, M. A. Uusitalo, B. Timus, and M. Fallgren, Scenarios for 5G mobile and wireless communications: The vision of the METIS project, *IEEE Communications Magazine*, vol. 52, no. 5, pp. 26–35, 2014.

3. A. Osseiran, J. F. Monserrat, and P. Marsch, *5G mobile and Wireless Communications Technology*, Cambridge University Press, New York, NY, United States, 2016.

4. F. Boccardi, R. W. Heath Jr, A. Lozano, T. L. Marzetta, and P. Popovski, Five disruptive technology directions for 5G, *IEEE Communications Magazine*, vol. 52, no. 2, pp. 74–80, 2014.

5. NGMN Alliance 5G white paper, 2015, https://www.ngmn.org/5g-whitepaper/5g-white-paper.html

6. F. Meshkati, H. V. Poor, S. C. Schwartz, and N. B. Mandayam, An energy-efficient approach to power control and receiver design in wireless data networks, *IEEE Transactions on Communications*, vol. 53, no. 11, pp. 1885–1894, 2005.

7. A. Janjic, L. Velimirovic, M. Stankovic, and A. Petrusic, Commercial electric vehicle fleet scheduling for secondary frequency control, *Electric Power Systems Research*, vol. 147, pp. 31–41, 2017.

8. A. Janjic, and L. Velimirovic, Optimal scheduling of utility electric vehicle fleet offering ancillary services, *ETRI Journal*, vol. 37, no. 2, pp. 273–282, 2015.

9. A. Janjic, L. Velimirovic, J. D. Velimirovic, and Z. Dzunic, Internet of Things in power distribution networks – state of the art, *52nd International Scientific Conference on Information, Communication and Energy Systems and Technologies*, Niš, Serbia, June 28–30, pp. 333–336, 2017.

10. A. Janjic, and L. Velimirovic, Integrated fault location and isolation strategy in distribution networks using Markov decision process, *Electric Power Systems Research*, vol. 180, pp. 1–9, 2020.

11. A. Janjic, S. Savic, L. Velimirovic, and V. Nikolic, Renewable energy integration in smart grids-multicriteria assessment using the Fuzzy analytical hierarchy process, *Turkish Journal of Electrical Engineering and Computer Sciences*, vol. 23, no. 6, pp. 1896–1912, 2015.

12. A. Janjic, and L. Velimirovic, Bivariate statistics of lightning density and guaranteed quality of service in distribution network using copulas, *Electric Power Systems Research*, vol. 194, 107059, 2021.

13. A. W. Scott, and R. Frobenius, Multiple access techniques: FDMA, TDMA, and CDMA, in *RF Measurements for Cellular Phones and Wireless Data Systems*, A. W. Scott, R. Frobenius, Ed. Wiley-IEEE Press, New York, NY, United States, pp. 413–429, 2008.

14. H. Li, G. Ru, Y. Kim, and H. Liu, OFDMA capacity analysis in MIMO channels, *IEEE Transactions on Information Theory*, vol. 56, no. 9, pp. 4438–4446, 2010.

15. S. M. R. Islam, N. Avazov, O. A. Dobre, and K.-S. Kwak, Power-domain non-orthogonal multiple access (NOMA) in 5G systems: Potentials and challenges, *IEEE Communications Surveys & Tutorials*, vol. 19, no. 2, pp. 721–742, 2017.

16. L. Dai, B. Wang, Y. Yuan, S. Han, I. Chih-Lin, and Z. Wang, Nonorthogonal multiple access for 5G: Solutions, challenges, opportunities, and future research trends, *IEEE Communications Magazine*, vol. 53, no. 9, pp. 74–81, 2015.

17. A. Benjebbour, K. Saito, A. Li, Y. Kishiyama, and T. Nakamura, Nonorthogonal multiple access (NOMA): Concept, performance evaluation and experimental trials, *Proceedings IEEE International Conference on Wireless Networks and Mobile Communications (IEEE WINCOM'15)*, Marrakech, Morocco, October 20–23, pp. 1–6, 2015.

18. Z. Ding, M. Peng, and H. V. Poor, Cooperative non-orthogonal multiple access in 5G systems, *IEEE Communications Letters*, vol. 19, no. 8, pp. 1462–1465, 2015.

19. D. Wan, M. Wen, F. Ji, H. Yu, and F. Chen, Non-orthogonal multiple access for cooperative communications: Challenges, opportunities, and trends, *IEEE Wireless Communications*, vol. 25, no. 2, pp. 109–117, 2018.

20. Z. Ding, Z. Yang, P. Fan, and H. V. Poor, On the performance of nonorthogonal multiple access in 5G systems with randomly deployed users, *IEEE Signal Processing Letters*, vol. 21, no. 12, pp. 1501–1505, 2014.

21. K. Kusume, G. Bauch, and W. Utschick, IDMA vs. CDMA: Analysis and comparison of two multiple access schemes, *IEEE Transactions on Wireless Communications*, vol. 11, no. 1, pp. 78–87, 2012.

22. D. Guo, and C.-C. Wang, Multiuser detection of sparsely spread CDMA, *IEEE Journal on Selected Areas in Communications*, vol. 26, no. 3, pp. 421–431, 2008.

23. M. Al-Imari, M. A. Imran, R. Tafazolli, and D. Chen, Subcarrier and power allocation for LDS-OFDM system, *Proceedings IEEE 73rd Vehicular Technology Conference (VTC Spring)*, Budapest, Hungary, May 15–18, pp. 1–5, 2011.

24. M. Taherzadeh, H. Nikopour, A. Bayesteh, and H. Baligh, SCMA codebook design, *Proceedings IEEE Vehicular Technology Conference (IEEE VTC'14 Fall)*, September, pp. 1–5, 2014.

25. L. Lu, Y. Chen, W. Guo, H. Yang, Y. Wu, and S. Xing, Prototype for 5G new air interface technology SCMA and performance evaluation, *China Communications*, vol. 12, Supplement, pp. 38–48, 2015.

26. X. Dai, S. Chen, S. Sun, S. Kang, Y. Wang, Z. Shen, and J. Xu, Successive interference cancelation amenable multiple access (SAMA) for future wireless communications, *Proceedings IEEE International Conference on Communication Systems (IEEE ICCS'14)*, Macau, China, November 19–21, pp. 1–5, 2014.

27. I. Chih-Lin, S. Han, Z. Xu, Q. Sun, and Z. Pan, 5G: rethink mobile communications for 2020+, *Philosophical Transactions of the Royal Society A: Mathematical, Physical and Engineering Sciences*, vol. *374*, no. 2062, 2016, pp. 1–13.

28. S. Kang, X. Dai, and B. Ren, Pattern division multiple access for 5G, *Telecommunications Network Technology*, vol. 5, no. 5, pp. 43–47, 2015.

29. J. Huang, K. Peng, C. Pan, F. Yang, and H. Jin, Scalable video broadcasting using bit division multiplexing, *IEEE Transactions on Broadcasting*, vol. 60, no. 4, pp. 701–706, 2014.

30. Z. Yuan, G. Yu, and W. Li, Multi-user shared access for 5G, *Telecommunications Network Technology*, vol. 5, no. 5, pp. 28–30, 2015.

31. I.C. Lin, J. Huang, R. Duan, C. Cui, J. Jiang, and L. Li, Recent Progress on C-RAN Centralization and Cloudification, *IEEE Access*, vol. 2, pp. 1030–1039, 2014.

32. L. Z. Velimirovic, and S. Maric, New adaptive compandor for LTE signal compression based on spline approximations, *ETRI Journal*, vol. 38, no. 3, pp. 463–468, 2016.

33. A. Benjebbour, A. Li, Y. Kishiyama, H. Jiang, and T. Nakamura, System-level performance of downlink NOMA combined with SU-MIMO for future LTE enhancements, In *Proceedings IEEE Globecom Workshops (GC Wkshps)*, Austin, TX, USA, December 08–12, pp. 706–710, 2014.

34. E. Paolini, C. Stefanovic, G. Liva, and P. Popovski, Coded random access: Applying codes on graphs to design random access protocols, *IEEE Communications Magazine*, vol. 53, no. 6, pp. 144–150, 2015.

35. B. Wang, K. Wang, Z. Lu, T. Xie, and J. Quan, Comparison study of non-orthogonal access schemes for 5G, *Proceedings IEEE International Symposium on Broadb and Multimedia Systems and Broadcasting*, Ghent, Belgium, June 17–19, pp. 1–5, 2015.

36. F. L. Luo, and C. Zhang, *Signal Processing for 5G: Algorithms and Implementations*, John Wiley and Sons, New York, NY, United States, 2016.

37. R. Hoshyar, F. P. Wathan, and R. Tafazolli, Novel low-density signature for synchronous CDMA systems over AWGN channel, *IEEE Transactions Signal Processing*, vol. 56, no. 4, pp. 1616–1626, 2008.

38. J. Van De Beek, and B. M. Popovic, Multiple access with low-density signatures, *GLOBECOM IEEE Global Telecommunications Conference*, Honolulu, Hawai, USA, November 30, pp. 1–6, 2009.

39. S. Golomb, and E. Posner, Rook domains, Latin squares, affine planes, and error-distributing codes, *IEEE Transactions on Information Theory*, vol. 10, no. 3, pp. 196–208, 1964.

40. X. Dai, Z. Zhang, B. Bai, S. Chen, and S. Sun, Pattern division multiple access: A new multiple access technology for 5G, *IEEE Wireless Communications*, vol. 25, no. 2, pp. 54–60, 2018.

41. Z. Ding, Y. Liu, J. Choi, Q. Sun, M. Elkashlan, I. Chih-Lin, and H. V. Poor, Application of non-orthogonal multiple access in LTE and 5G networks, *IEEE Communications Magazine*, vol. 55, no. 2, pp. 185–191, 2017.

42. O. Moreno and A. Tirkel, Algebraic generators of sequences for communication signals, US Patent 9413421 B2, 2016.

43. M. R. Schroeder, *Number Theory in Science and Communications*, New York: Springer, 1997.

44. M. Su, and A. Winterhof, Autocorrelation of Legendre–Sidelnikov Sequences, *IEEE Transactions on Information Theory*, vol. 56, no. 4, pp. 1714–1718, 2010.

45. T. Koshy, *Elementary Number Theory with Applications*, Academic Press, San Diego, CA, United States, 2002.

46. V. M. Sidelnikov, Some k-valued pseudo-random sequences and nearly equidistant codes, *Problemy Peredachi Informatsii*, vol. 5, no. 1, pp. 12–16, 1969.

47. A. Lempel, M. Cohn, and W. Eastman, A class of balanced binary sequences with optimal autocorrelation property, *IEEE Transactions on Information Theory*, vol. IT-23, no. 1, pp. 38–42, 1977.

48. Z. Guohua, and Z. Quan, Pseudonoise codes constructed by Legendre sequence, *Electronics Letters*, vol. 38, no. 8, pp. 376–377, 2002.

49. Y.-S. Kim, J.-S. Chung, J.-S. No, and H. Chung, New families of M-ary sequences with low correlation constructed from Sidelnikov sequences, *IEEE Transactions on Information Theory*, vol. 54, no. 8, pp. 3768–3774, 2008.

50. J. J. Rushanan, Weil sequences: A family of binary sequences with good correlation properties, *Proceedings IEEE International Symposium on Information Theory (ISIT2006)*, Seattle, WA, USA, July 9–14, pp. 1648–1652, 2006.

51. Y. K. Han, and K. Yang, New M-ary sequence families with low correlation and large size, *IEEE Transactions on Information Theory*, vol. 55, no. 4, pp. 1815–1823, 2009.

52. N. Y. Yu, and G. Gong, Multiplicative characters, the Weil bound, and polyphase sequence families with low correlation, *IEEE Transactions on Information Theory*, vol. 56, no. 12, pp. 6376–6387, 2010.

53. S. W. Golomb, *Shift Register Sequences*, Laguna Hills, CA: Aegean Park, 1982.

54. C. Ding, T. Helleseth, and W. Shan, On the linear complexity of Legendre sequences, *IEEE Transactions on Information Theory*, vol. 44, no. 3, pp. 1276–1278, 1998.

55. S. W. Golomb, Algebraic constructions for Costas arrays, *Journal of Combinatorial Theory, Series A*, vol. 37, no. 1, pp. 13–21, 1984.

56. S.W. Golomb, and H. Taylor, Constructions and properties of Costas arrays, *Proceedings of the IEEE*, vol. 72, no. 9, pp. 1143–1163, 1984.

57. J. P. Costas, Medium constraints on sonar design and performance, Tech. Rep. Class 1 Rep. R65EMH33, General Electric Company, Fairfield, CT, USA, November 1965.

58. J. P. Costas, A study of a class of detection waveforms having nearly ideal range-Doppler ambiguity properties, *Proceedings of the IEEE*, vol. 72, no. 8, pp. 996–1009, 1984.

59. S. V. Maric, I. Seskar, and E. L. Titlebaum, On cross-ambiguity properties of Welch-Costas arrays, *IEEE Transactions on Aerospace and Electronic Systems*, vol. 30, no. 4, pp. 1063–1071, 1994.

60. S. V. Maric, and E. L. Titlebaum, A class of frequency hop codes with nearly ideal characteristics for use in multiple-access spread-spectrum communications and radar and sonar systems, *IEEE Transactions on Communications*, vol. 40, no. 9, pp. 1442–1447, 1992.

61. H. D. Luke, Sequences and arrays with perfect periodic correlation, *IEEE Transaction on Aerospeace and Electronic Systems*, vol. 24, no. 3, pp. 287–294, 1988.

62. O. Moreno, and S.V. Maric, A new family of frequency-hop codes, *IEEE Transactions on Communications*, vol. 48, no. 8, pp. 1241–1244, 2000.

63. R. G. van Schyndel, A. Z. Tirkel, I. D. Svalbe, T. E. Hall, and C. F. Osborne, Algebraic construction of a new class of Quasi-orthogonal arrays in steganography, *Security and Watermarking of Multimedia Contents*, vol. 3657, pp. 354–364, 1999.

8 Ambient Backscatter Communication

A Solution for Energy-Efficient 5G-Enabled IoT

Tushar S. Muratkar

Indian Institute of Information Technology, Nagpur, India

Ankit Bhurane and Ashwin Kothari

Visvesvaraya National Institute of Technology, Nagpur, India

Robin Singh Bhadoria

GLA University, Mathura, India

CONTENTS

DOI: 10.1201/9781003045809-10

8.1   INTRODUCTION

All cellular technologies from first (1G) to fourth generation (4G) have made an immense contribution to societal development. Fifth generation (5G) is a key player in the Internet of Things (IoT) which provides seamless connectivity to numerous smart devices such as sensors, washing machines, TVs, tablets and smartphones. 5G offers a variety of advantages for IoT that are not present on 4G or other technologies. These include the capacity of 5G to serve a large number of smart machines with a variety of speed, bandwidth and quality of service requirements. For the efficient working of IoT, smart devices should be capable of fast data transfer, and must be low powered and have low latency. However, fast data transfer requires more energy, draining the batteries of connected devices more quickly. Frequent recharging and replacing of batteries is a huge task especially in the case of environmental sensors located in high-risk areas. The IoT, therefore, demands devices that consume less power, have low cost and are less bulky. One of the solutions to this is the ambient backscatter communication system (ABCS). ABCS works on the principle of backscattering the already existing ambient radio frequency (RF) signals, e.g., radio, television, wireless fidelity (Wi-Fi), etc. The first work on backscatter communication (BackCom) was done in 1948 [1]. Here the author demonstrated was was described as "Communication by Means of Reflected Power" and used it to identify the target in the sight of radar.

8.1.1   TYPES OF BACKCOM SYSTEMS

BackCom systems are of three types, depending on the configuration of entities: i) Monostatic BackCom System; ii) Bi-static BackCom System; and iii) BackCom Ambient System. All three configurations consist of a tag and a reader as primary components.

8.1.2   MONOSTATIC BACKCOM SYSTEM (MBCS)

The MBCS configuration (Figure 8.1) consists of a tag and a reader. Here the backscatter receiver is positioned within the reader itself. The reader acts as a carrier emitter (RF source) and is used to activate the tag. Once the tag is activated, it modulates the incoming RF signals and backscatters them towards the RF source that captures these backscattered signals. The MBCS is used in radio frequency identification (RFID) small-distance communication. MBCS suffers from two major limitations. First, the backscattered signal experiences round-trip path loss as the carrier emitter and backscatter receiver are co-positioned at the reader [2]. Second, it suffers from a double near-far

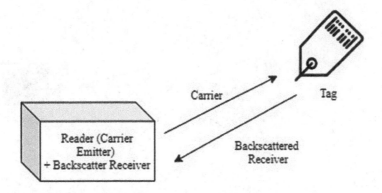

FIGURE 8.1   Configuration of monostatic BackCom system.

effect, where users close to the carrier emitter harvest maximum energy and hence generate better performance than users further away from the RF source [3].

8.1.2.1   Bistatic BackCom System (BBCS)

The BBCS configuration (Figure 8.2) differs from the MBCS in the sense that in BBCS, the carrier emitter and a receiver of backscatter are not located at the same reader. Separation of backscatter receiver and RF source helps to overcome the disadvantage of round-trip path loss and double near-far effect. Thus in BBCS, the RF source is flexible from the location point of view and this flexibility helps improve the communication range compared with MBCS. Another advantage of BBCS is its simple design, which reduces manufacturing costs of carrier emitter and backscatter receiver as compared to MBCS [4].

8.1.2.2   Ambient BackCom System (ABCS)

The topology of BackCom is almost the same as BBCS with one significant difference. Unlike BBCS, ABCS does not use a dedicated RF source; instead, already existing RF sources in the near vicinity (e.g., TV tower, radio tower, Wi-Fi access point, etc.) are used. In this case, the tag harvests the energy as well as information from the ambient RF source, then it modulates its information, and

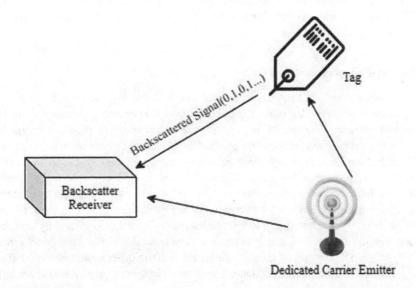

FIGURE 8.2   Configuration of bistatic BackCom system.

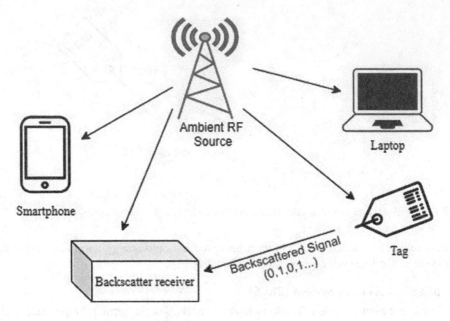

FIGURE 8.3   Configuration of ambient BackCom system.

finally, it backscatters the bits (either '0' or '1') towards the receiver (Figure 8.3). The ABCS configuration offers numerous advantages over MBCS and BBCS due to its ability to effectively utilize already existing ambient RF sources for communication. Since dedicated RF sources do not need to be maintained, the cost and power consumption of ABCS is low. It revamps spectrum resource utilization because it uses already existing ambient RF sources so no new frequency spectrum allocation is needed. One downside of ABCS is that it has less stability and consistency than BBCS, as it uses ambient RF signals that are continuously changing.

8.1.3 OVERVIEW OF BACKCOM SYSTEMS

Despite their different configurations, the basic principle of all three types of BackCom system is the same. The key elements of BackCom are the backscatter transmitter (tag) and the backscatter receiver (reader).

8.1.4 BACKSCATTER TRANSMITTER

The tag used in BackCom is either active or passive. A dedicated power source or a battery is implanted inside the active tag. A passive tag has no power source and is powered using an incident RF signal coming from an external RF source. Figure 8.4. represents the backscatter transmitter. When the incoming signals are incident on to the tag, it undergoes antenna impedance matching and backscatters (reflects) the signals to the backscatter receiver (reader). Generally, the amount of reflection (backscattering) depends on the difference in the impedance values. To control the rate of backscattering, the backscatter transmitter consists of a switch that changes the antenna impedance, ultimately changing the amount of energy reflected by the antenna. The input to the switch is a sequence of '1' and '0' bits, and the switch includes a transistor that is internally coupled with the antenna. Whenever the transistor input is zero, it is deactivated and the impedances are matched; consequently a very small amount of signal is reflected. On the other hand, whenever the transistor input is one, it is activated and the impedances are mismatched, as a result of which a large amount

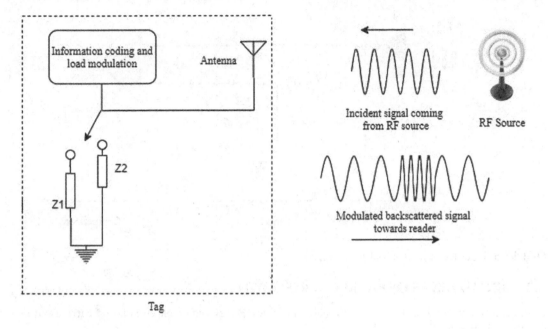

FIGURE 8.4 Backscatter transmitter.

of signal is reflected. Thus, the presence of one and zero bits at the transistor controls the rate of backscattering by the tag and conveys the information to the backscatter receiver. The impedance mismatch between antenna and chip of the tag is represented by the reflection coefficient, Γ, and this reflection coefficient of the tag is given as [5]

$$\Gamma_j = \frac{Z_j - Z_t^*}{Z_j + Z_t} \tag{8.1}$$

where Z_t denotes impedance of antenna, Z_t^* its complex conjugate version and $j = 1, 2$ denotes toggling state.

8.1.5 BACKSCATTER RECEIVER

Designing a receiver for BackCom is a challenging task. Two major challenges are: i) the ambient RF signal information is already encoded in different forms, hence piggybacking the backscattered signal over it makes the overall process more difficult; and ii) decoding of the backscattered signal must not require a large amount of power. Power-consuming elements such as analog-to-digital converter (ADC) and oscillators are therefore contraindicated. To overcome these challenges, authors in [6] designed the detector at the backscatter receiver (Figure 8.5) using only ultra-low power analog components. Here the first stage is an averaging stage, the second is the thresholding stage, followed finally by a comparator stage. The input to the average envelope detection stage is the RF signal. This stage averages the variations in the RF signal. The second stage performs threshold computation between the two levels. Lastly, a comparator circuit compares the average envelope signal with the threshold and generates the bits. In this manner, an RF signal with natural variations is converted into bits using low-power analog components.

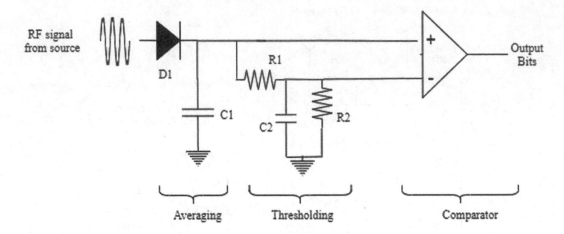

FIGURE 8.5 Detector used in backscatter receiver.

8.2   BROAD AREAS OF BACKCOM RESEARCH

In the following section, the various domains in which BackCom research is being carried out are discussed in detail [7].

8.2.1   Signal Processing

The signal processing domain in BackCom deals with various aspects such as channel coding techniques, interference effects, channel decoding techniques and signal detection.

8.2.1.1   Channel Coding

The purpose of channel coding techniques is to provide every possible support for data recovery. The channel coding techniques used in conventional wireless communication involve greater complexity and use high levels of power, making them unsuitable for low-power IoT applications. To overcome this problem, the Electronic Product Code (EPC) enables the EPC Gen-2 communication protocol, which provides two different encoding techniques suitable for BackCom: (i) Miller code and (ii) Bi-Phase space (FM0) code. These encoding schemes have certain merits such as reduced noise, improved reliability of signal and simplicity. NRZ and Manchester coding techniques are also popular in MBCS. A major disadvantage of the NRZ and Manchester coding techniques is that NRZ code includes a long length of '1' or '0' bits and the Manchester code needs more bits than the actual signal. Authors in [8] demonstrated the use of 6/8 balanced block code to the backscatter link and increased throughput by at least 50% with less cost, reduced complexity and low power consumption. The communication range of BBCS is increased by the use of cyclic code-based short-block length cyclic error-correcting codes [9]. Numerical observations indicate that this method achieves a range of 150m and consumes 20mW of power. Parks et al. [10] proposed a new coding mechanism named μcode that provides the advantages of code division multiple-access (CDMA) technology. With this new coding scheme, the authors achieved a higher data rate and larger communication range. With μcode the tags can exchange information over a distance of more than 80 feet. Tao et al. [11] presented the concept of Manchester code and differential Manchester code-based BackCom systems. In Manchester code '0' and '1' are mapped to either '01' or '10'. The differential Manchester code is designed based on simple logic. Whenever there is a change between the current bit and preceding bit, it is encoded by '1', otherwise by '0'. In [12], the authors presented a new encoding scheme, orthogonal space-time block code (OSTBC), for a dyadic backscatter channel with multiple antennas at tag and reader. The main advantage of OSTBC is that it achieves maximum diversity order by keeping the decoding complexity linear.

8.2.1.2 Interference

In ABCS the tag leverages the ambient RF signals and performs the BackCom. However, in practice, the ambient RF signals may collide with legal receivers. In order to trace the exact position of bit collision, the authors in [13] presented an anti-collision algorithmic rule for Miller subcarrier and FM0 code. Simulation results indicate that the throughput is increased from 0.35 to 0.5 in the case of the FM0 code as well as the Miller code. 5G-enabled IoT will involve communication between billions of devices daily. As a result, there is a huge likelihood of interference between these devices. Sabharwal et al. [14] presented the in-band full-duplex (IBFD) technique, which leads to higher throughput but also causes self-interference. Interference in MBCS is divided into two types: (i) interference between the readers (R2R); and (ii) interference between reader and tag (R2T) [15]. R2R interference leads to incorrect detection of the signal, more bandwidth usage and security risks, while R2T interference also leads to tag jamming. In [16], the authors investigated BER performance due to the interference of ambient RF signals on the legacy devices located in the surrounding. Simulation results indicate that BER is affected by angle and the distance between the tags.

8.2.1.3 Channel Decoding

Channel decoding plays an important role in BackCom, as the backscattered signals are low power and hence make information extraction a critical job. However, in the literature, there exist numerous channel decoding techniques enabling information to be extracted from low-powered reflected signals. Authors in [17] proposed a new decoding rule called composite hypothesis testing that is capable of achieving high diversity order in a bistatic scatter radio network. Accurate channel estimation and strict synchronization between the tags are needed when decoding concurrent tag transmissions. To cater to these constraints, in [18] authors introduced the BiGroup which is a new RFID communication pattern for commodity-off-the-shelf (COTS) RFID tags that can detect the collision of data packets from COTS tags by not following the above-mentioned constraints. Also in that work, no synchronization is needed for COTS tags to amalgamate with the continuing communications. Most of the concurrent decoding schemes in the literature consider that signals reflected from tag are very stable; however, this fails in the case of dynamic backscatter systems. In order to deal with dynamic backscatter systems, authors in [19] introduced a parallel decoding technique termed Flip Tracer that achieves a throughput of about 2 Mbps.

8.2.1.4 Signal Detection

Signal detection in BackCom is categorized as coherent and non-coherent detection. Coherent detection (CD) differs from non-coherent detection (NCD) in that coherent detection requires signal phase information while performing detection and employs matched filters to determine the data. Coherent detection in BackCom is more challenging than conventional wireless communication, as BackCom involves three channels (source to tag, source to the reader, and tag to the reader), unlike single channels in conventional wireless communication. The various frameworks related to the antenna such as its scattering efficiency, structural mode and reflection coefficients add further complexity to the coherent detection process. As the phase of the signal varies in phase-shift keying (PSK) modulation, coherent detection is preferred. Hilliard et al. [9] employed coherent detection for BBCS. Experimental results indicate that with the proposed method, the authors achieved theoretical gains of 18 dB.

In contrast to CD, the NCD does not require signal phase information, so amplitude shift keying (ASK) and frequency shift keying (FSK) are preferred. Qian et al. [20] investigated non-coherent detection in ABCS with unknown channel state information (CSI). Here authors derived the maximum likelihood detector by considering the joint probability density function (PDF) of received signal vectors. They also designed a joint energy detector that waives the requirement for prior knowledge of the ambient signals. The detection process in BackCom requires threshold estimation which is a time-consuming process, so the authors in [11] developed semi-coherent Manchester (SeCoMC) and non-coherentManchester (NoCoMC) detectors that put an end to the estimation of decision threshold and allow instant symbol-by-symbol detection. In [21], the authors carried out a theoretical study of semi-coherent detection in ABCS. Here instead of channel estimation, the pilot symbols are transmitted to obtain the parameters required in the detection process.

8.2.2 BackCom: Wireless Communications

In the following subsections, we discuss different types of modulation techniques and multiple access techniques for the signals that are backscattered from the tag.

8.2.2.1 Modulation

For efficient data transmission, various modulation techniques such as ASK, FSK and PSK are proposed in the literature. Of these three, PSK is widely used in ABCS, while ASK and FSK are extensively used in BBCS. The simplest version of ASK is on-off keying (OOK). Authors in [22] presented two different modulation techniques, OOK and FSK, for BBCS. FSK is more advantageous than OOK in terms of increased receiver sensitivity, wider range and simpler multiple access. These modulation techniques have low data rate, i.e., they transfer only one bit per symbol period, and hence a higher-order modulation technique consisting of M states is used in [23]. In this case, a hardware model is designed and with 4-PSK modulation, they achieved a data rate of 20 kbps for a range of 2.5 feet. Similarly, Correia et al. [24] presented the use of the M-quadrature amplitude modulation (M-QAM) technique to support higher data rates. They built a modulator containing a Wilkinson power divider and two transistors that act as a switch and generate the M-QAM. The major problem with M-QAM modulation is that it is easily affected by noise, leading to normalized power loss. Orthogonal frequency-division multiplexing (OFDM) is extensively used in digital television, audio broadcasting and 4G communications, etc. Yang et al. [25] carried out a theoretical study on OFDM-based ABCS. The system model is based on the concept of spread spectrum modulation. The authors studied the proposed model for both single-antenna and multiple-antenna systems. Simulation results show that there is a significant improvement in BER performance with the use of multiple antennas at the receiver. Authors in [26] presented full-duplex MBCS, in which the carrier emitter and commercial Bluetooth chipset are co-located. The Bluetooth low energy (BLE) uses a Gaussian-shaped Binary Frequency Shift Keying (GFSK) modulation scheme that works at 1 Mbps. In total there are 40 BLE channels and for every channel whenever there is positive frequency deviation of more than 185 kHz above the channel center frequency, a '1' bit is represented and a '0' bit is represented by negative frequency deviation of more than 185 kHz below the channel center frequency.

8.2.2.2 Multiple Access Techniques in BackCom Systems

To support a large number of users at a time, BackCom uses multiple access strategies based on time division, frequency division and code division. User cooperation is an important deciding factor when choosing various types of multiple access techniques. In [26], the authors discussed user cooperation in a wireless powered network. In this, the system model consists of two users that use time division multiple access (TDMA) to uplink information and one hybrid access point (H-AP). One of the users is located nearer to the H-AP, and hence enjoys good channel conditions as well as needing less time to uplink its data towards H-AP. As a part of user cooperation, the nearer user dedicates some part of its uplink time and harvested energy to conveying the far user's information to the H-AP. Many BackCom protocols follow the TDMA approach, which is sequential, and hence do not support concurrent transmissions of the signal. To overcome this, authors in [27] introduced a backscatter spike train (BST) approach that enables parallel transmission by using intra-bit multiplexing of OOK signals from the multiple tags. The BST detects the collided bits by detecting the signal edges using a high-rate sampling backscatter reader. The results indicate that the performance of the BST technique is 10 times higher than that of the normal TDMA approach. Liu et al. [28] carried out a study on a backscatter multiplicative multiple access systems (BM-MAC). Unlike conventional linear additive multiple access systems, here the backscatter transmitter performs multiplication operations. As all the devices in MBCS make use of a common channel, it is affected by the tag collision problem. While the TDMA approach can be used to overcome this limitation, this approach has several inherent limitations itself, such as multipath distortion and already defined time slots. The EPC global generation 2 proposed a fixed spectrum allocation (FSA)-based CDMA algorithm to solve the tag collision problem [29].

8.2.3 BackCom: Wireless Information and Power Transfer

Conventional RF signals include information as well as energy simultaneously. Earlier the information and energy were considered to be separate from one another. But now researchers are uniting energy and radio wave information so that the RF spectrum is used efficiently. Due to extensive improvements in wireless power transfer (WPT), present-day wireless technologies such as mm-waves, 5G and small cells include WPT [30]. As regards energy harvesting from non-dedicated RF sources, authors in [31] surveyed the city of London using four harvesters, including antenna, impedance-matching circuit, rectifier, maximum power point tracking and storage elements. The experimental results indicate that out of a total of 270 Underground stations, more than 50% are suitable to be deployed as ambient energy harvesters. Also, in urban (semi-urban) conditions, single-band harvesters are 40% efficient and require less power of about -25dBm. A wireless energy transfer (WET)-enabled BackCom model consisting of active nodes (AN) and passive nodes (PN) is discussed in [32]. The authors developed a harvest-while-scatter protocol (HWS) in which the AN emits RF waves continuously and the PN first harvests the RF energy, later modulates their data and lastly scatters the modulated wave towards the receiver. The main advantage of integrating data transfer and energy harvesting is that PN has a long time to harvest and hence can have more RF energy. A WPT to the $0.01mm^2$ CMOS tag coil was demonstrated in [33]. The tag consumes $0.1mW$ DC power and operates at 4.7GHz. The proposed model uses OOK backscattering and achieves a data rate of 20kbps. A simultaneous wireless information power transfer (SWIPT)-based decode and forward (DF) relay-enabled ABCS is presented in [34], where energy harvesting is carried out using power-splitting relaying protocol. The SWIPT technique has limitations of non-linearity, sensitivity and saturation of farfield energy harvesting. Authors in [35] presented a piecewise linear approximation model minimizing the above problems in the SWIPT technique. The observations indicate that SWIPT research should consider non-linearity and limited sensitivity of the harvester.

8.2.4 Task Scheduling and Resource Allocation

The working of the BackCom system is decided by the effective use of available resources and their scheduling. Task scheduling considers the energy constraints and compatibility of the tasks. In a traditional wireless-powered communication network, the only RF energy harvesting is carried out during the first time slot and this leads to its underutilization, since no information transfer takes place during the said time slot. To overcome this problem, authors in [36] proposed a backscatter-assisted wireless-powered communication network (BAWPCN) in which the initial time slot is for backscattering, and harvest-then-transfer (HTT) is used in the remaining time slots. IoT involves multiple sensors used for a variety of applications such as environment monitoring, vehicle networking, body area sensor networks and real-time tracking, etc., and so the biggest problem in such cases is how to collect data from these various sensors. Also, different sensors may have different data priorities, i.e., urgent or normal. Therefore in [37], the authors presented a data-scheduling mechanism that takes into account both priorities and status of the data. The Markov decision process (MDP) is also used, which optimizes the cost associated with the scheduling mechanism. The overall life of a wireless network improves drastically with the proper allocation of available resources (bandwidth, power, etc.). To maximize the fairness of wireless-powered cooperative cognitive radio networks, Kalamkar et al. [38] presented three resource allocation schemes: i) equal time allocation;

ii) proportional time allocation; and iii) minimum throughput maximization. The fairest of these schemes is minimum throughput maximization, but it has the least secondary-user sum throughput. Lyu et al. [39] investigated resource allocation in multi-user BackCom consisting of a reader and multiple tags. They considered three resources, time allocation to the passive and semi-passive tags, reader transmission power and reflection ratio. Also, authors in [40] observed that optimal time and energy allocation schemes improve overall efficiency in BackCom.

8.3 MATHEMATICAL ASPECTS OF ABCS

A typical ABCS consists of a three-node elementary model consisting of an ambient RF source, a passive tag and a reader (Figure 8.6). [21]. Let us assume that all the channels are frequency flat and block fading [41], where all of these channels remain unchanged for a particular channel coherence time but may vary independently in different coherence time intervals. Let the channel coefficients between source to reader, source to tag and tag to reader be represented by h, f, and g, respectively. The ambient RF source emits its RF signal which can reach both tag and reader. This RF signal can be considered to have complex Gaussian distribution or phase shift keying (PSK) [42], or it can be an orthogonal frequency-division multiplexing (OFDM) signal [43]. The tag modulates the received RF signal by changing the load impedance of its antenna, and thus backscatters the modulated signal towards the reader. Whenever there is impedance matching, the maximum amount of RF signal is absorbed; consequently a very small amount of signal is reflected and this represents a non-back-scattering state (bit '0'), whereas whenever there is impedance mismatching, very little RF signal is absorbed and a large amount of signal is reflected, representing a backscattering state (bit '1'). The signal received by the tag can be represented mathematically as

$$x[n] = fs[n], \tag{8.2}$$

where $s[n]$ is the signal broadcast by the RF source. As the tag consists of only passive components, the thermal noise at the tag is neglected, as is the case in most of the literature [22, 44, 45].

We know that the tag backscatters the signal in binary form, and so let $b[n] \in 0,1$ denote the backscattered signal by the tag, where the '0' and '1' are transmitted with equal probabilities. Thus signal backscattered from the tag is given as

$$x_b[n] = \eta b[n] x[n], \tag{8.3}$$

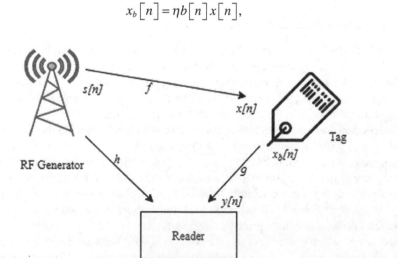

FIGURE 8.6 Three-node elementary model of ABCS.

where η is the reflection coefficient of the tag, which is usually a complex number with absolute value less than 1 [46].

The signals from both RF source and backscatter transmitter are received at the reader and this received signal is represented as

$$y[n] = hs[n] + gx_b[n] + w[n], \tag{8.4}$$

By making some simple mathematical manipulations, the above equation can be rewritten as

$$y[n] = (h + \eta fgb[n])s[n] + w[n], \tag{8.5}$$

where $w[n]$ is additive white Gaussian noise (AWGN) having zero mean and variance as N_w i.e., $w[n] \sim CN(0, N_w)$. From Equation (8.4), we aim to detect the desired signal; however, detecting it in the presence of interfering signal as well as noise is a challenging task. A proper detector therefore needs to be designed at the reader. Various detectors are designed in the literature, e.g., maximum likelihood detector [11], energy detector [47], covariance-based detector [48], differential coded maximum likelihood detection [2], maximum a posteriori (MAP) detector, etc.

To understand the detection process, let us consider the case of a maximum likelihood detector at the reader. In order to make the reader capable of separating the desired signal from an interfering one, the rate of backscatter is much lower than the rate of ambient signals. Thus we consider that $b[n]$ do not change for N consecutive $s[n]$s where N is an even number. For simplicity, let us assume that b denotes one symbol backscattered by the tag. Hence, based on whether the b is '0' or '1', the signal received at the reader is given by

$$y[n] = \begin{cases} h_0 s[n] + w[n], b = 0 \\ h_1 s[n] + w[n], b = 1 \end{cases} \tag{8.6}$$

where $h_0 = h$ and $h_1 = h + \eta fg$. If we consider that the signal broadcast by the RF source has complex Gaussian distribution, i.e., $s[n] \sim CN(0, P_s)$, the received signal vector also follows complex Gaussian distribution. Hence, corresponding to the backscattered symbol, the received signal can also be expressed as

$$y[n] = \begin{cases} \sim CN(0, \sigma_0^2 I_N), b = 0 \\ \sim CN(0, \sigma_1^2 I_N), b = 1 \end{cases} \tag{8.7}$$

where $\sigma_0^2 \triangleq |h_0|^2 P_s + N_w$, $\sigma_1^2 \triangleq |h_1|^2 P_s + N_w$ and I_N is the unit vector of order N. The aim of the maximum likelihood detector is to minimize the bit error rate (BER) and for this, the decision can be made by comparing two likelihood functions. The likelihood ratio test is defined as [49]

$$\Lambda(y) = \frac{p(y|H_0)}{p(y|H_1)} = \left(\frac{\sigma_0^2}{\sigma_1^2}\right)^N \exp\left(\frac{\sigma_0^2 - \sigma_1^2}{\sigma_0^2 \sigma_1^2} Z\right) \tag{8.8}$$

where $p(y|H_i)$ denotes probability density function (PDF) and $Z = \|y\|^2$. Finally a threshold is obtained and by comparing Z with threshold, the decision whether '0' or '1' is received is made.

8.4 UPCOMING BACKSCATTER COMMUNICATION TECHNIQUES

In this section, the various upcoming BackCom techniques are discussed.

8.4.1 VISIBLE LIGHT BACKCOM SYSTEMS (VLBCS)

The VLBCS is extremely useful in areas where RF-assisted BackCom is not possible e.g., hospitals or airplanes. Li et al. [50] developed the prototype of a duplex VLBCS that is capable of performing bi-directional communication. Specifically, the system consists of an LED lamp (ViReader) and a passive tag (ViTag). ViTag consists of an LCD modulator and retro-reflector, and it communicates by reflecting and modulating the incoming light. This system achieves a data rate of 10 kbps in the downlink and 0.5 kbps in uplink over a distance of up to 2.4 m. The use of on-off keying (OOK) modulation in VLBCS affects throughput. To overcome this problem, authors in [51] proposed and implemented pixelated VLBCS using 8 PAM (pulse amplitude modulation). The prototype is composed of numerous LCD shutters and small reflectors that form the number of pixels. With 8 PAM the data rate is three times that of the OOK modulation technique at a distance of 2 meters. Experimental results indicate that a data rate of 600 bps is achieved for a distance of 2 meters. This data rate can be further increased by using a faster modulator in place of the LCD shutter.

Xu et al. [52] proposed trend-based modulation and a code-assisted demodulation technique, using which they built a battery-free ViTag that achieves an uplink data rate of up to 1 kbps. The authors analyzed performance of their VLBCS in terms of success probability and network capacity using stochastic geometry [53]. Specifically, the network topology is modeled using a generalized Gauss-Poisson process. They also studied the impact of the duty cycle and reflection coefficient on system performance, and observed that a larger value of the duty cycle causes more interference and hence decreases the success probability. Also, the reflection coefficient should have an appropriate value. This is because too small and too large values of the reflection coefficient result in reduced received power at the receiver and larger interference power, respectively. Authors in [54] built an ambient light BackCom system that is capable of backscattering the data at a rate of 100 bps for 10 cm distance.

8.4.2 RELAY-ASSISTED BACKCOM SYSTEM

Using relay leads to an increase in the range of the communication system. In a typical relay-based system, the relay receives the signal from the transmitter and re-transmits it towards the receiver. The advantage of a relay-based system is that it reduces the BER with the same transmission distance between source and destination. To this end, Yan et al. [55] analyzed performance of a two-way decode and forward relay-based ABCS. The closed-form expression of BER was obtained and results indicate that this relay-based model surpasses the direct transmission system. Authors in [56] studied time-switching relay-assisted BackCom. The two separate cases for a relay are considered: a relay with and without an embedded energy source. When the relay has an embedded energy source, it forwards the information using its energy and in the case of absence of an embedded energy source, it first harvests the energy from ambient RF signals, then uses the harvested energy to forward the information. Researchers in [57] presented a decode-and-forward hybrid relay scheme that demonstrates both ambient backscattering and wireless-powered communications mode. For coordination between these two processes, a mode selection protocol is proposed that works according to network conditions. Jia and Zhou [58] examined a decode-and-forward relay scheme in which the backscatter device itself works as a relay. Test statistics were obtained for on-off keying modulated signals at both the relay and the destination (receiver). Simulation results indicate that prime obstruction in BER performance is due to relay to the destination link. To overcome this problem, the source must supply maximum power to the relay.

Researchers in [59] presented a two-hop backscatter relay model and devised a through-put maximization problem that optimizes both relay strategies and wireless power transfer. Simulated results indicate that relay strategy leads to significant improvement in throughput, especially when the power demand of the radio is low. To improve the communication range of battery-less IoT networks, the authors in [60] came up with the concept of using drones as relays, termed 'RFly'. RFly is full duplex and is capable of preserving the phase as well as timing characteristics of the forwarded data packets. A hardware prototype of RFly capable of communicating between 50 and 55 m with an average accuracy of 19 cm has been built. Due to the growing demand for unmanned aerial vehicles (UAVs) in wireless communication, Yang et al. [61] examined the use of UAVs in BackCom. Here, the proposed network involves several backscatter devices (BDs) and carrier emitters (CEs) mounted on the ground and UAVs. The major limitation of such a system is that there is limited energy available with UAV and carrier emitter. These problems are overcome by jointly optimizing UAV trajectory, BD scheduling, and CE transmission power.

8.4.3 mm-Wave-Based BackCom

The millimeter-wave (mm-wave) is a part of the 5G wireless communication standard that makes a significant contribution to boosting its data rate. The 5G spectrum covers the frequencies in the range of sub-6 GHz to 100 GHz, with the mm-wave spectrum ranging from 24 GHz to 100 GHz. Due to extensive use of lower frequencies with TV, radio signals and existing 4G networks, they are heavily congested, resulting in slower data speeds. Unlike these lower frequencies, the mm-wave spectrum is relatively unused, so it offers high bandwidth and hence a very high data rate. This mm-wave technology can cover smaller areas with a very high-speed data rate, unlike the lower frequencies that cover large areas but at lower speeds. Due to the capability of mm-wave to offer high speed in smaller areas, it finds applications in ABCS. Authors in [62] built the first hardware model of mm-wave BackCom and achieved a data rate of 4Gbps. The proposed system works in the range of 24–28 GHz with an energy consumption of less than 0.15 pJ/bit.

8.4.4 Long Range (Lo-Ra) BackCom

Traditional BackCom is known for its low-power, low-cost and short-range communication. Increase in range is possible but at the cost of high power consumption. However, [63] designed a Lo-Ra BackCom system that can effectively communicate over a distance of 475 m. The Lo-Ra is based on chirp-spread spectrum (CSS) modulation, in which a '0' bit is denoted as a continuous chirp that undergoes linear increase with, and a '1' bit is a chirp that undergoes cyclical shifting in time.

8.4.5 Ultra-Wide Band (UWB) BackCom

Generally speaking, most ambient backscattering devices are designed for a certain RF source, for instance, a TV tower or an FM tower. But, in other cases, signals from a specific source may not be available and/or may be unstable. To solve this problem, the backscatter technique for the UWB is introduced in [64]. Here, instead of depending on a single ambient RF source, the UWB backscatter device takes the signal from multiple ambient RF emitters and exchanges the information by reflecting these RF signals towards the universal backscatter receiver, as illustrated in Figure 8.7. This way of implementing the BackCom system has a certain number of advantages, including:

1. There is less dependency on a single ambient RF source and hence the system can work effectively even when some of the ambient RF sources are unreachable.
2. It results in an improvement in signal-to-noise ratio (SNR), so such a BackCom system can also be deployed in remote geographical areas.

3. The increased value of SNR also improves the overall communication range of the system. Experimental results from [64] indicate that by backscattering the signals from 17 ambient RF sources, an overall communication range of up to 50 m with 1 kbps data rate can be achieved.

However, designing a UWB BackCom system is a very challenging task, especially the universal backscatter receiver. This is because the receiver is receiving reflected signals from a number of RF sources, so the channels between them may have different variations. Hence the most challenging task is designing a detector at the receiver that is capable of decoding the received signals with their dissimilar characteristics. To overcome this challenge, in [64], a wideband antenna that receives reflected as well as ambient RF source signals is used at the universal backscatter receiver. Further, all these received signals are allowed to pass through a bandpass filter, are down-converted, and finally undergo maximal ratio combining. Although the UWB BackCom system offers the above-mentioned advantages, it is not energy efficient and so further research needs to be done.

8.4.6 FULL-DUPLEX BACKCOM

The amalgamation of full duplexing with ABCS was first presented by [65]. In this, the backscatter transmitter communicates with the backscatter receiver at the same time and same frequency as the backscatter receiver sends feedback to backscatter transmitter, although at a lower rate. Feedback information plays a very important role in increasing the efficiency of overall ABCS. This is because, based on the feedback from the backscatter receiver, if there is a collision of packets, a backscatter transmitter can halt its transmission. Also, the receiver can feed back about BER, enabling the transmitter to immediately adjust its backscatter rate accordingly. In addition, the feedback information helps the transmitter to send only those bits that are errors instead of resending the entire data packets.

Designing full-duplex ABCS is challenging because, at the time of backscattering feedback information, there is a large change in received signal amplitude which ultimately affects the decoding capability of the backscatter receiver. To overcome this problem, authors in [65] developed a novel scheme in which the backscatter receiver absorbs and feeds back a fixed number of signals so that there is no change in the amplitude of received signals. Authors in [66] investigated the resource allocation for full-duplex ABCS with ambient OFDM signals. By jointly optimizing the allocated time, power reflection coefficients and subcarrier power allocation of the backscatter devices, the performance of all the backscatter devices is maximized. In [67], the authors studied a full-duplex BackCom technique with symbiotic radio (SR), with particular reference to structure and power consumption. They also derived the expression for achievable rates of backscatter devices with Gaussian and other modulation codewords.

8.4.7 COGNITIVE RADIO NETWORK (CRN) WITH BACKCOM

Nowadays, due to the extensive use of wireless technologies, the spectrum is scarce. Efficient use of the available spectrum is achieved with CRN. CRN can automatically detect unused channels in the spectrum and change the transmission parameters of the cognitive devices, enabling more communication to run in parallel and also improving radio operating behavior. Specifically, the cognitive devices look out for spectrum gaps to transmit data. A spectrum gap is nothing but a frequency band which is already allocated to licensed users but is somehow unused. Unlicensed users may access the spectrum gap. Thus, we can say that a CRN is an autonomous, adaptive and multi-dimensional system that decides its future action by learning from its own experiences. In traditional RF-powered CRN, the secondary transmitter (ST) first extracts the energy from primary signals and then, based on the extracted energy, transmits its data to the secondary receiver (SR). However, in this approach, the performance of ST largely depends upon total energy harvested and the duration that the primary channel is busy or idle. To solve this issue, [68] combined CRN with ambient BackCom, in which the ST not only performs energy harvesting from primary signals but also backscatters the ambient RF signals to the SR and thus, in general, the throughput of the secondary system is enhanced.

Energy efficiency or minimization of energy consumption is an important metric from the perspective of RF-powered ambient backscatter CRN. It is the ratio of the average achievable throughput to the average energy consumption, measured in bits/Hz/J [69]. The energy efficiency is affected during spectrum sensing. Authors in [70] carried out a study of sensing errors in energy-efficient ambient backscatter CRN. They derived analytical expressions for average energy consumption, average throughput and energy efficiency. In [71], the authors analyzed the performance of relay-based ambient backscatter CRN, deriving the exact closed-form expressions in terms of outage probability and throughput.

8.4.8 Non-Orthogonal Multiple Access (NOMA) in BackCom

The multiple access methods used in 1G to 4G allow multiple users to access wireless networks. The common feature of these multiple access methods is that they orthogonalize different users' signals by allocating different resources such as time, frequency, space or code to various users. Orthogonality restricts the number of users. 5G wireless standards must be able to support a large number of users, and have high throughput and low latency. These requirements can be more efficiently fullfilled by NOMA than with traditional orthogonal multiple access. Owing to this, [72] studied the relationship between IoT and cellular networks, presented the concept of backscatter NOMA in integration with symbiotic, and analyzed outage probability and ergodic rates. In [73], the authors studied MBCS using NOMA, where power-domain NOMA is implemented. To maximize system performance, the authors suggested design guidelines for reflection coefficients. Simulation results indicate that under less severe channel conditions, system performance can be enhanced to a great extent by setting proper reflection coefficients. The authors in [74] proposed an unmanned aerial vehicle (UAV)-assisted NOMA BackCom network to spot the trade-off between backscatter coefficients, the total number of backscatter devices and UAV altitude. The main objective of the study was to reduce UAV flight time and increase the number of successfully decoded bits. Yang et al. [75] studied resource allocation problems for NOMA-based BackCom networks. The system model consists of multiple single-antenna backscatter devices (BD) and a single multi-antenna backscatter receiver (BR). The BDs receive energy from the carrier emitter and reflect them towards the BR. To enhance the spectrum efficiency, a new transmission protocol is proposed, which integrates dynamic time-division multiple-access (TDMA) and NOMA.

8.5 APPLICATIONS OF BACKCOM

8.5.1 BackCom in Medical Science

The sensors used in medical science for the treatment of patients need less power and involves low computing cost. In these conditions, the sensors communicate using ambient Wi-Fi signals. Abdelnasser et al. [76] designed a Wi-Fi-based gesture recognition system for humans that uses changes in the Wi-Fi signal strength to identify hand motions with the help of reflected signals. They also developed a prototype of this system using laptops and evaluated it using standard Wi-Fi access points (APs). Experimental results indicate that the system has an accuracy of 87.5% using single AP, and the accuracy increases to 96 % with three APs. The accuracy remains robust even with human interference. Authors in [77] use RF signals reflected from the person's body to detect emotions (happiness, sadness, etc). Thus, there is no need for a person to carry sensors all over their body. To trace the emotions exactly, it is necessary to measure the minute variations of individual beat length. This is made possible using a newly designed algorithm which has achieved accuracy similar to that of on-body ECG monitors. Moradi et al. [78] use the BackCom technique to communicate with the implanted biomedical sensors needed in brain–machine interface (BMI). Specifically, the authors

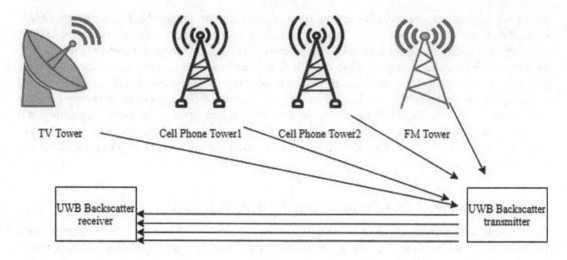

FIGURE 8.7   UWB BackCom system.

designed a 1x1x1 mm³ backscattering device (connected to RFID IC) capable of receiving sufficient wireless power to activate the IC. Researchers in [79] presented backscatter designs especially for deep-tissue devices. Deep-tissue systems are affected by two major complications, surface interference and signal deflection. The signal deflection occurs as the RF signal moves eight times slower in muscles than in air, and as a result, a large portion of the signal is reflected. To solve these problems, the authors designed a novel backscatter technique termed Remix. This technique gets rid of interference from the body surface and also locates the implanted backscatter devices. The numerical observations indicate that the proposed model can work with an accuracy of 1.4 cm in animal tissue and can effectively communicate with an average SNR of 15.2 dB at 1 MHz bandwidth.

Authors of [63] used CSS-modulated Lo-Ra BackCom for communicating with contact lens and patch sensors. Here the backscatter device sends LoRa packets at 1 MHz across a 104 ft by 32 ft atrium. Experimental results show that reliable connectivity can be maintained when the received signal strength is above -132 dBm. Iyer et al. [80] introduced a novel inter-technology BackCom for use in implanted devices as well as a smart contact lens. Inter-technology BackCom means wireless transmissions are transformed from one form to another over the air. In particular, Wi-Fi and Zig-Bee signals are created using Bluetooth transmission. In this case, the authors built a prototype and showed that without dedicated reader hardware, the smart contact lens can exchange information with Wi-Fi and Bluetooth radios. The prototype consists of a 1cm-diameter 30 AWG antenna that is enclosed in a polydimethylsiloxane (PDMS) thin film. When the distance between the prototype and Wi-Fi receiver is 10 inches the received power is up to -74 dBm, and in the case of a Bluetooth receiver it is up to -82 dBm. Zhang et al. [81] presented a working model based on frequency-shifted backscatter for on-body sensors that shifts the carrier signal towards adjacent non-overlapping Wi-Fi or Bluetooth bands. This model can operate up to 4.8m with a data rate ranging from a few bps to tens of kbps and power consumption of 45μW. The data rate can be further increased depending on the number of transmitters and receivers; a data rate of up to 48.7 kbps is achieved when two transmitters and receivers are taken into consideration.

8.5.2   BACKCOM FOR SMART CITIES/SMART HOMES

Energy-efficient BackCom has a vital role to play in creating smart cities and smart homes that improve our quality of life. A smart home involves multiple passive sensors located at various locations including ceiling, furniture, walls, etc. These tags are powered using ambient RF sources (e.g., TV towers, Wi-Fi AP, cellular stations, FM towers, etc.). The sensors perform various tasks like detection of smoke and gas leaks, indoor positioning, movement monitoring and surveillance [82]. Using the BackCom technique, simple household objects are also made smart. For example, smart dustbins have been created that can keep an eye on their trash levels and by backscattering can communicate the information to passing garbage trucks. In the case of smart cities, the BackCom sensors can be positioned at different locations (trees, buildings, parking lights, street lights, etc.). These sensors keep track of pollution and traffic levels in our environment, providing us with improved safety and quality of life. Again, in this case these sensors can be powered using ambient RF sources.

8.5.3 BackCom in Smart Factories

BackCom sensors are widely used in smart factories due to their low power budget and reduced cost. Typical work in this sector involves the movement of cargo from one place to another. In this scenario, BackCom tags help gather the information even without lifting heavy objects. In this sector, one of the emerging techniques is BackCom-based three-dimensional orientation tracking, which consists of an array of tags that are mounted on the surface of a target object and provide reader three-dimensional information by exploring the relative phase difference between different tags.

8.5.4 BackCom in Precision Agriculture

The use of BackCom in agriculture can boost productivity to a considerable extent. Precision agriculture means the use of smart devices at the right time, right place and in the right manner to achieve improved profitability, sustainability and protection of resources. To this end, authors in [83] developed a BackCom-based leaf-sensing mechanism. The system consists of a tag attached to a temperature leaf sensor with power consumption of about 20μW. Wireless communication was established using Morse code modulation.

8.6 OPEN RESEARCH ISSUES

The concept of ABCS is in its developing stages. For proper use of IoT in 5G wireless standards, device battery life has to be increased proportionally. In this respect, ABCS is the perfect solution. The main focus of this section is to discuss open research issues related to the effective use of ABCS in 5G to create an ecosystem for IoT.

8.6.1 Interference Management

The prime issue of interference is with legal communication systems. The tag in ABCS uses carriers from nearby RF sources (legal devices) and backscatters them towards the reader. Thus while backscattering, there should not be any interference with these legacy receivers. Liu et al. [6] observed that the backscattering tag does not affect nearby legacy devices for a data rate of fewer than 10 kbps. However, currently, the BackComnetworks are operating at higher data rates and in such cases this may affect nearby legacy devices. Such interference can be tracked using appropriate stochastic geometry models and spatial analysis.

Interference also occurs with other BackCom systems that are present at the same location and work on the same frequency. This can be solved using interference cancellation and interference alignment schemes for particular channels [84]. To deal with double fading channels, Zhang et al. [85] proposed a multiple tag selection and combining scheme in which a BackCom wireless channel was correlated with Nakagami-fading channels.

8.6.2 PHYSICAL LAYER SECURITY

ABCS are mainly deployed in an environment that consists of multiple tiny sensors exchanging information with one another. They may carry confidential data and hence BackCom mechanisms need to be highly secure. These sensors have limited computational power, therefore it is not feasible to use highly complex algorithms and authentication protocols. To solve this problem, people in [86] proposed an artificial noise-aided tag-scheduling scheme in which, initially, the best tag is selected based on channel gain between tag and reader, and later another tag is selected that generates artificial noise that affects an eavesdropper. However, this approach results in high energy consumption and increases co-channel interference with the adjacent user. A physical layer security scheme for maximizing the secrecy rate of MBCS was presented by [87], in which along with conventional signals the reader broadcasts noise towards the eavesdropper.

8.6.3 MACHINE-LEARNING ALGORITHMS

With the tremendous success of machine-learning (ML) algorithms in various scientific domains like image processing, researchers are now aiming to use ML in wireless communication. The use of ML in wireless networks can significantly increase the overall management and hence the productivity of a communication system. The advantage of using ML is that it can solve complex problems without explicit programming. Also, the use of ML algorithms protects BackCom devices from privacy vulnerabilities. The detectors used in conventional ABCS suffer from high bit error rate and hence [88] makes use of ML that transforms the detection problem into a classification problem. Hong et al. [89] presented an ML-based antenna design scheme that is capable of achieving directional communication between tag and reader. Simulation results indicate that it can maintain the quality of the main channel and prevents information leakage. Wang et al. [90] in their work used classification algorithms to detect the binary PSK modulated information with an acceptable performance without knowledge of channel state information (CSI) and the constellations of the legacy system. Here, the authors tested their work with a simple modulation scheme; however, in the future it can also be extended to higher-order modulation schemes as well as multi-tag scenarios.

8.6.4 ACHIEVING HIGH DATA RATES

The BackCom is a low-rate communication system. For example, [6] achieves a data rate of 1 kbps. Effective use of IoT with 5G communications requires a high data rate, but achieving this rate leads to excessive power consumption, making this a challenging task.

8.7 CONCLUSION

The ambient BackCom system is an effective solution that takes into account the need for energy-efficient IoT in which sensors and miniature devices exchange information with a strict emphasis on less computing and a low power budget. Since ABCS does not need huge batteries, it can be used to garner the benefits of 5G to expand IoT networks. In this chapter we have presented the ABCS as a solution to energy-efficient 5G-enabled IoT. First, different types of BackCom systems were discussed along with their configurations. Then, the fundamentals of the BackCom system in terms of transmitter and receiver sides were explained. Further discussion included the broad areas where BackCom research is carried out and the mathematical aspects of ABCS. Finally we presented upcoming BackCom techniques and its applications, and highlighted the open research issues for further exploration.

REFERENCES

1. H. Stockman, "Communication by means of reflected power," *Proceedings of the IRE*, vol. 36, no. 10, pp. 1196–1204, Oct. 1948.
2. G. Wang, F. Gao, R. Fan, and C. Tellambura, "Ambient backscatter communication systems: Detection and performance analysis," *IEEE Transactions on Communications*, vol. 64, no. 11, pp. 4836–4846, Nov. 2016.
3. D. S. Choi, "Backscatter radio communication for wireless powered communication networks," in *2015 21st Asia-Pacific Conference on Communications (APCC)*, Oct. 2015, pp. 370–374.
4. X. Lu, D. Niyato, H. Jiang, D. I. Kim, Y. Xiao, and Z. Han, "Ambient backscatter networking: A novel paradigm to assist wireless powered communications," *arXiv preprint arXiv:1709.09615*, 2017.
5. J. Zhang, G. Y. Tian, M. Marindra, A. I. Sunny, and A. B. Zhao, "A review of passive rfid tag antenna-based sensors and systems for structural health monitoring applications," *Sensors*, vol. 17, no. 2, p. 265, Feb. 2017.
6. V. Liu, A. Parks, V. Talla, S. Gollakota, D. Wetherall, and J. R. Smith, "Ambient backscatter: Wireless communication out of thin air," in *ACM SIGCOMM Computer Communication Review*, vol. 43, no. 4, pp. 39–50, Aug. 2013.
7. T. S. Muratkar, A. Bhurane, and A. Kothari, "Battery-less internet of things–a survey," *Computer Networks*, vol. 180, p. 107385, 2020.
8. G. D. Durgin and B. P. Degnan, "Improved channel coding for next-generation rfid," *IEEE Journal of Radio Frequency Identification*, vol. 1, no. 1, pp. 68–74, 2017.
9. N. Fasarakis-Hilliard, P. N. Alevizos, and A. Bletsas, "Coherent detection and channel coding for bistatic scatter radio sensor networking," *IEEE Transactions on Communications*, vol. 63, no. 5, pp. 1798–1810, 2015.
10. A. N. Parks, A. Liu, S. Gollakota, and J. R. Smith, "Turbocharging ambient backscatter communication," *ACM SIGCOMM Computer Communication Review*, vol. 44, no. 4, pp. 619–630, 2014.
11. Q. Tao, C. Zhong, H. Lin, and Z. Zhang, "Symbol detection of ambient backscatter systems with manchester coding," *IEEE Transactions on Wireless Communications*, vol. 17, no. 6, pp. 4028–4038, 2018.
12. C. Boyer and S. Roy, "Space time coding for backscatter rfid," *IEEE Transactions on Wireless Communications*, vol. 12, no. 5, pp. 2272–2280, Mar. 2013.
13. Y. G. Kim and H. Vinck, "Anticollision algorithms for fm0 code and miller subcarrier sequence in rfid applications," *IEEE Transactions on Vehicular Technology*, vol. 67, no. 6, pp. 5168–5173, Jun. 2018.
14. A. Sabharwal, P. Schniter, D. Guo, D. W. Bliss, S. Rangarajan, and R. Wichman, "In-band full-duplex wireless: Challenges and opportunities," *IEEE Journal on selected areas in communications*, vol. 32, no. 9, pp. 1637–1652, 2014.
15. A. Bekkali, S. Zou, A. Kadri, M. Crisp, and R. Penty, "Performance analysis of passive uhf rfid systems under cascaded fading channels and interference effects," *IEEE Transactions on Wireless Communications*, vol. 14, no. 3, pp. 1421–1433, 2014.
16. C. Chen, G. Wang, Y. Wang, and Q. Miao, "Interference analysis of ambient backscatter on existing wireless communication systems," in *2017 IEEE 85th Vehicular Technology Conference (VTC Spring)*, Jun. 2017, pp. 1–5.
17. P. N. Alevizos and A. Bletsos, "Noncoherent composite hypothesis testing receivers for extended range bistatic scatter radio wsns," in *2015 IEEE International Conference on Communications (ICC)*, Jun. 2015, pp. 4448–4453.
18. J. Ou, M. Li, and Y. Zheng, "Come and be served: Parallel decoding for cots rfid tags," in *Proceedings of the 21st Annual International Conference on Mobile Computing and Networking*, Sep. 2015, pp. 500–511.
19. M. Jin, Y. He, X. Meng, Y. Zheng, D. Fang, and X. Chen, "Fliptracer: Practical parallel decoding for backscatter communication," *IEEE/ACM Transactions on Networking (TON)*, vol. 27, no. 1, pp. 330–343, 2019.
20. J. Qian, F. Gao, G. Wang, S. Jin, and H. Zhu, "Noncoherent detections for ambient backscatter system," *IEEE Transactions on Wireless Communications*, vol. 16, no. 3, pp. 1412–1422, 2016.
21. J. Qian, F. Gao, G. Wang, S. Jin, and H. Zhu, "Semi-coherent detection and performance analysis for ambient backscatter system," *IEEE Transactions on Communications*, vol. 65, no. 12, pp. 5266–5279, 2017.

22. J. Kimionis, A. Bletsas, and J. N. Sahalos, "Increased range bistatic scatter radio," *IEEE Transactions on Communications*, vol. 62, no. 3, pp. 1091–1104, 2014.

23. J. Qian, A. N. Parks, J. R. Smith, F. Gao, and S. Jin, "Iot communications with m-psk modulated ambient backscatter: Algorithm, analysis, and implementation," *IEEE Internet of Things Journal*, vol. 6, no. 1, pp. 844–855, 2018.

24. R. Correia, A. Boaventura, and N. B. Carvalho, "Quadrature amplitude backscatter modulator for passive wireless sensors in iot applications," *IEEE Transactions on Microwave Theory and Techniques*, vol. 65, no. 4, pp. 1103–1110, 2017.

25. G. Yang, Y.-C. Liang, R. Zhang, and Y. Pei, "Modulation in the air: Backscatter communication over ambient ofdm carrier," *IEEE Transactions on Communications*, vol. 66, no. 3, pp. 1219–1233, 2017.

26. H. Ju and R. Zhang, "User cooperation in wireless powered communication networks," in *2014 IEEE Global Communications Conference*, 2014, pp. 1430–1435.

27. P. Hu, P. Zhang, and D. Ganesan, "Leveraging interleaved signal edges for concurrent backscatter," in *Proceedings of the 1st ACM Workshop on Hot Topics in Wireless*, 2014, pp. 13–18.

28. W. Liu, Y.-C. Liang, Y. Li, and B. Vucetic, "Backscatter multiplicative multiple-access systems: Fundamental limits and practical design," *IEEE Transactions on Wireless Communications*, vol. 17, no. 9, pp. 5713–5728, 2018.

29. W. Wei, J. Su, H. Song, H. Wang, and X. Fan, "Cdma-based anti-collision algorithm for epc global c1 gen2 systems," *Telecommunication Systems*, vol. 67, no. 1, pp. 63–71, 2018.

30. K. Huang and X. Zhou, "Cutting the last wires for mobile communications by microwave power transfer," *IEEE Communications Magazine*, vol. 53, no. 6, pp. 86–93, 2015.

31. M. Pinuela, P. D. Mitcheson, and S. Lucyszyn, "Ambient rf energy harvesting in urban and semi-urban environments," *IEEE Transactions on microwave theory and techniques*, vol. 61, no. 7, pp. 2715–2726, 2013.

32. Q. Yao, A. Huang, H. Shan, and T. Q. Quek, "Wet-enabled passive communication networks: Robust energy minimization with uncertain csi distribution," *IEEE Transactions on Wireless Communications*, vol. 17, no. 1, pp. 282–295, 2017.

33. N.-C. Kuo, B. Zhao, and A. M. Niknejad, "Inductive wireless power transfer and uplink design for a cmos tag with 0.01 mm 2 coil size," *IEEE Microwave and Wireless Components Letters*, vol. 26, no. 10, pp. 852–854, 2016.

34. S. T. Shah, K. W. Choi, T.-J. Lee, and M. Y. Chung, "Outage probability and throughput analysis of swipt enabled cognitive relay network with ambient backscatter," *IEEE Internet of Things Journal*, vol. 5, no. 4, pp. 3198–3208, 2018.

35. P. N. Alevizos and A. Bletsas, "Sensitive and nonlinear far-field rf energy harvesting in wireless communications," *IEEE Transactions on Wireless Communications*, vol. 17, no. 6, pp. 3670–3685, 2018.

36. B. Lyu, Z. Yang, G. Gui, and Y. Feng, "Throughput maximization in backscatter assisted wireless powered communication networks," *IEICE Transactions on Fundamentals of Electronics, Communications and Computer Sciences*, vol. 100, no. 6, pp. 1353–1357, 2017.

37. D. T. Hoang, D. Niyato, P. Wang, D. I. Kim, and L. B. Le, "Optimal data scheduling and admission control for backscatter sensor networks," *IEEE Transactions on Communications*, vol. 65, no. 5, pp. 2062–2077, 2017.

38. S. S. Kalamkar, J. P. Jeyaraj, A. Banerjee, and K. Rajawat, "Resource allocation and fairness in wireless powered cooperative cognitive radio networks," *IEEE Transactions on Communications*, vol. 64, no. 8, pp. 3246–3261, 2016.

39. B. Lyu, Z. Yang, G. Gui, and Y. Feng, "Optimal resource allocation policies for multi-user backscatter communication systems," *Sensors*, vol. 16, no. 12, pp. 1–14, 2016.

40. R. Wang and D. R. Brown, "Throughput maximization in wireless powered communication networks with energy saving," in *2014 48th Asilomar Conference on Signals, Systems and Computers*, 2014, pp. 516–520.

41. J. Qian, F. Gao, G. Wang, S. Jin, and H. Zhu, "Noncoherent detections for ambient backscatter system," *IEEE Transactions on Wireless Communications*, vol. 16, no. 3, pp. 1412–1422, 2016.

42. J. Qian, F. Gao, G. Wang, S. Jin, and H. Zhu, "Semi-coherent detector of ambient backscatter communication for the internet of things," in *2017 IEEE 18th International Workshop on Signal Processing Advances in Wireless Communications (SPAWC)*, 2017, pp. 1–5.

43. G. Yang, Q. Zhang, and Y.-C. Liang, "Cooperative ambient backscatter communications for green internet-of-things," *IEEE Internet of Things Journal*, vol. 5, no. 2, pp. 1116–1130, 2018.

44. C. Boyer and S. Roy, "Invited paper—backscatter communication and rfid: Coding, energy, and mimo analysis," *IEEE Transactions on Communications*, vol. 62, no. 3, pp. 770–785, 2013.

45. G. Yang, C. K. Ho, and Y. L. Guan, "Multi-antenna wireless energy transfer for backscatter communication systems," *IEEE Journal on Selected Areas in Communications*, vol. 33, no. 12, pp. 2974–2987, 2015.

46. Q. Tao, C. Zhong, K. Huang, X. Chen, and Z. Zhang, "Ambient backscatter communication systems with mfsk modulation," *IEEE Transactions on Wireless Communications*, vol. 18, no. 5, pp. 2553–2564, 2019.

47. M. A. ElMossallamy, M. Pan, R. Jantti, K. G. Seddik, G. Y. Li, and Z. Han, "Noncoherent backscatter communications over ambient ofdm signals," *IEEE Transactions on Communications*, vol. 67, no. 5, pp. 3597–3611, 2019.

48. Y. Liu, G. Wang, Z. Dou, and Z. Zhong, "Coding and detection schemes for ambient backscatter communication systems," *IEEE Access*, vol. 5, pp. 4947–4953, 2017.

49. J. Proakis, *Digital Communications*, 5th edition. McGraw-Hill, 2007.

50. J. Li, A. Liu, G. Shen, L. Li, C. Sun, and F. Zhao, "Retro-vlc: Enabling battery-free duplex visible light communication for mobile and iot applications," in *Proceedings of the 16th International Workshop on Mobile Computing Systems and Applications*, 2015, pp. 21–26.

51. S. Shao, A. Khreishah, and H. Elgala, "Pixelated vlc-backscattering for self-charging indoor iot devices," *IEEE Photonics Technology Letters*, vol. 29, no. 2, pp. 177–180, 2016.

52. X. Xu, Y. Shen, J. Yang, C. Xu, G. Shen, G. Chen, and Y. Ni, "Passivevlc: Enabling practical visible light backscatter communication for battery-free iot applications," in *Proceedings of the 23rd Annual International Conference on Mobile Computing and Networking*, 2017, pp. 180–192.

53. X. Wang, K. Han, and M. Zhang, "Modeling the large-scale visible light backscatter communication network," in *2017 23rd Asia-Pacific Conference on Communications (APCC)*, 2017, pp. 1–6.

54. J. Yun and B.-J. Jang, "Ambient light backscatter communication for iot applications," *Journal of Electromagnetic Engineering and Science*, vol. 16, no. 4, pp. 214–218, 2016.

55. W. Yan, L. Li, G. He, X. Li, A. Gao, H. Zhang, and Z. Han, "Performance analysis of two-way relay system based on ambient backscatter," in *2018 13th IEEE Conference on Industrial Electronics and Applications (ICIEA)*, 2018, pp. 1853–1858.

56. B. Lyu, Z. Yang, T. Xie, G. Gui, and F. Adachi, "Optimal time allocation in relay assisted backscatter communication systems," in *2018 IEEE 87th Vehicular Technology Conference (VTC Spring)*, 2018, pp. 1–5.

57. X. Lu, D. Niyato, H. Jiang, E. Hossain, and P. Wang, "Ambient backscatter-assisted wireless-powered relaying," *IEEE Transactions on Green Communications and Networking*, vol. 3, no. 4, pp. 1087–1105, 2019.

58. X. Jia and X. Zhou, "Decode-and-forward relaying using a backscatter device: Power allocation and ber analysis," in *2019 IEEE Global Communications Conference (GLOBE-COM)*, 2019, pp. 1–6.

59. S. Gong, J. Xu, L. Gao, X. Huang, and W. Liu, "Passive relaying scheme via backscatter communications in cooperative wireless networks," in *2018 IEEE Wireless Communications and Networking Conference (WCNC)*, 2018, pp. 1–6.

60. Y. Ma, N. Selby, and F. Adib, "Drone relays for battery-free networks," in *Proceedings of the Conference of the ACM Special Interest Group on Data Communication*, 2017, pp. 335–347.

61. G. Yang, R. Dai, and Y.-C. Liang, "Energy-efficient uav backscatter communication with joint trajectory and resource optimization," in *ICC 2019–2019 IEEE International Conference on Communications (ICC)*, 2019, pp. 1–6.

62. J. Kimionis, A. Georgiadis, and M. M. Tentzeris, "Millimeter-wave backscatter: A quantum leap for gigabit communication, rf sensing, and wearables," in *2017 IEEE MTT-S International Microwave Symposium (IMS)*, 2017, pp. 812–815.

63. V. Talla, M. Hessar, B. Kellogg, A. Najafi, J. R. Smith, and S. Gollakota, "Lora backscatter: Enabling the vision of ubiquitous connectivity," *Proceedings of the ACM on Interactive, Mobile, Wearable and Ubiquitous Technologies*, vol. 1, no. 3, pp. 1–24, 2017.

64. C. Yang, J. Gummeson, and A. Sample, "Riding the airways: Ultra-wideband ambient backscatter via commercial broadcast systems," in *IEEE INFOCOM 2017-IEEE Conference on Computer Communications*, 2017, pp. 1–9.

65. V. Liu, V. Talla, and S. Gollakota, "Enabling instantaneous feedback with full-duplex backscatter," in *Proceedings of the 20th Annual International Conference on Mobile Computing and Networking*, 2014, pp. 67–78.

66. G. Yang, D. Yuan, and Y.-C. Liang, "Optimal resource allocation in full-duplex ambient backscatter communication networks for green iot," in *2018 IEEE Global Communications Conference (GLOBECOM)*, 2018, pp. 1–6.

67. R. Long, H. Guo, L. Zhang, and Y.-C. Liang, "Full-duplex backscatter communications in symbiotic radio systems," *IEEE Access*, vol. 7, pp. 21 597–21 608, 2019.

68. D. T. Hoang, D. Niyato, P. Wang, D. I. Kim, and Z. Han, "Ambient backscatter: A new approach to improve network performance for rf-powered cognitive radio networks," *IEEE Transactions on Communications*, vol. 65, no. 9, pp. 3659–3674, 2017.

69. S. Althunibat, M. Di Renzo, and F. Granelli, "Towards energy-efficient cooperative spectrum sensing for cognitive radio networks: An overview," *Telecommunication Systems*, vol. 59, no. 1, pp. 77–91, 2015.

70. R. Kishore, S. Gurugopinath, P. C. Sofotasios, S. Muhaidat, and N. Al-Dhahir, "Opportunistic ambient backscatter communication in rf-powered cognitive radio networks," *IEEE Transactions on Cognitive Communications and Networking*, vol. 5, no. 2, pp. 413–426, 2019.

71. D.-T. Do, T.-L. Nguyen, and B. M. Lee, "Performance analysis of cognitive relay-assisted ambient backscatter with mrc over nakagami-m fading channels," *Sensors*, vol. 20, no. 12, p. 3447, 2020.

72. Q. Zhang, L. Zhang, Y.-C. Liang, and P.-Y. Kam, "Backscatter-noma: A symbiotic system of cellular and internet-of-things networks," *IEEE Access*, vol. 7, pp. 20 000–20 013, 2019.

73. J. Guo, X. Zhou, S. Durrani, and H. Yanikomeroglu, "Backscatter communications with noma," in *2018 15th International Symposium on Wireless Communication Systems (ISWCS)*, 2018, pp. 1–5.

74. A. Farajzadeh, O. Ercetin, and H. Yanikomeroglu, "Uav data collection over noma backscatter networks: Uav altitude and trajectory optimization," in *ICC 2019–2019 IEEE International Conference on Communications (ICC)*, 2019, pp. 1–7.

75. G. Yang, X. Xu, and Y.-C. Liang, "Resource allocation in noma-enhanced backscatter communication networks for wireless powered iot," *IEEE Wireless Communications Letters*, vol. 9, no. 1, pp. 117–120, 2019.

76. H. Abdelnasser, M. Youssef, and K. A. Harras, "Wigest: A ubiquitous wifi-based gesture recognition system," in *2015 IEEE Conference on Computer Communications (INFO-COM)*, 2015, pp. 1472–1480.

77. M. Zhao, F. Adib, and D. Katabi, "Emotion recognition using wireless signals," in *Proceedings of the 22nd Annual International Conference on Mobile Computing and Networking*, 2016, pp. 95–108.

78. E. Moradi, S. Amendola, T. Bjorninen, L. Sydanheimo, J. M. Carmena, J. M. Rabaey, and L. Ukkonen, "Backscattering neural tags for wireless brain-machine interface systems," *IEEE Transactions on Antennas and Propagation*, vol. 63, no. 2, pp. 719–726, 2014.

79. D. Vasisht, G. Zhang, O. Abari, H.-M. Lu, J. Flanz, and D. Katabi, "In-body backscatter communication and localization," in *Proceedings of the 2018 Conference of the ACM Special Interest Group on Data Communication*, 2018, pp. 132–146.

80. V. Iyer, V. Talla, B. Kellogg, S. Gollakota, and J. Smith, "Inter-technology backscatter: Towards internet connectivity for implanted devices," in *Proceedings of the 2016 ACM SIGCOMM Conference*, 2016, pp. 356–369.

81. P. Zhang, M. Rostami, P. Hu, and D. Ganesan, "Enabling practical backscatter communication for on-body sensors," in *Proceedings of the 2016 ACM SIGCOMM Conference*, 2016, pp. 370–383.

82. C. Gao, Y. Li, and X. Zhang, "Livetag: Sensing human-object interaction through passive chipless wifi tags," in *15th USENIX Symposium on Networked Systems Design and Implementation (NSDI 18)*, 2018, pp. 533–546.

83. S. N. Daskalakis, G. Goussetis, S. D. Assimonis, M. M. Tentzeris, and A. Georgiadis, "A uw backscatter-morse-leaf sensor for low-power agricultural wireless sensor networks," *IEEE Sensors Journal*, vol. 18, no. 19, pp. 7889–7898, 2018.

84. O. Bello and S. Zeadally, "Intelligent device-to-device communication in the internet of things," *IEEE Systems Journal*, vol. 10, no. 3, pp. 1172–1182, 2014.

85. Y. Zhang, F. Gao, L. Fan, S. Jin, and H. Zhu, "Performance analysis for tag selection in backscatter communication systems over nakagamim fading channels," in *2018 IEEE International Conference on Communications (ICC)*, 2018, pp. 1–5.

86. J. Y. Han, J. Kim, and S. M. Kim, "Physical layer security improvement using artificial noise-aided tag scheduling in ambient backscatter communication systems," in *2019 Eleventh International Conference on Ubiquitous and Future Networks (ICUFN)*, 2019, pp. 432–436.

87. W. Saad, X. Zhou, Z. Han, and H. V. Poor, "On the physical layer security of backscatter wireless systems," *IEEE transactions on wireless communications*, vol. 13, no. 6, pp. 3442–3451, 2014.

88. Y. Hu, P. Wang, Z. Lin, M. Ding, and Y.-C. Liang, "Machine learning based signal detection for ambient backscatter communications," in *ICC 2019–2019 IEEE International Conference on Communications (ICC)*, 2019, pp. 1–6.

89. T. Hong, C. Liu, and M. Kadoch, "Machine learning based antenna design for physical layer security in ambient backscatter communications," *Wireless Communications and Mobile Computing*, vol. 2019, pp. 1–10, 2019.

90. X. Wang, R. Duan, H. Yigitler, E. Menta, and R. Jantti, "Machine learning-assisted detection for bpsk-modulated ambient backscatter communication systems," in *2019 IEEE Global Communications Conference (GLOBECOM)*, 2019, pp. 1–6.

9 Deployment and Analysis of Random Walk and Random Waypoint Mobility Model for WSN-Assisted IoT Hierarchical Framework

Anurag Shukla
NIT Raipur, Raipur, India

Sarsij Tripathi
MNNIT Allahabad, Allahabad, India

Malay Kumar and Arun Chauhan
School of Computer Science & Engineering, University of Petroleum and Energy Studies (UPES), Dehradun, India

CONTENTS

9.1 INTRODUCTION

The Sensor Network-Assisted Internet of Things (WSN-assisted IoT) is a community of various computing devices, sensors and smart objects. The participating devices interconnect with each other and exchange data via the internet. IoT applications can be programmed and equipped with a sensor to make them more intelligent; a sensor can trigger output when certain real-time conditions are met and make it more effective [1]. Information relating to any participating devices and data collected by the sensors can be accessed easily and very efficiently [2].

It is reported that worldwide, around 7 billion computing devices are already linked to the internet. By 2030, this number will have grown to 100 billion [3]. The rapid increase in devices, leading to explosive growth in the rate of data transfer [4], is attracting the attention of the research community to opportunities, challenges and possible solutions.

WSN-assisted IoT acts as a bridge between the digital and real worlds, each of which has its own responsibilities.. The network is deployed in the real world over a vast area and data is collected from sensors in digital form. Applying machine learning models to the collected digital data makes applications smarter and more effective. The actuator or sensor nodes in the network are responsible for sensing. The collected sense data needs to transmit to the base station (BS) or sink. The sensor nodes have battery energy sources, which cannot be recharged or replaced. The enlarged area of WSN-assisted IoT networks requires more nodes to provide coverage, resulting in a lengthier communication path from the sensor node to the BS, more energy use and reduced network lifetime [5].

Recent work has been reported to increase network lifetime using hierarchical routing schemes for WSN-assisted IoT [6–9]. Here we focus on addressing the challenges of such networks and propose a hierarchical framework for efficient energy consumption, clustering and routing.

According to the literature [10, 11], IoT is the superset in which IoT nodes are directly linked to the internet. In the case of WSN, the devices (sensors) may or may not be connected to the internet. WSN-assisted IoT networks have a very large number of sensor nodes that may or may not be directly connected to the internet. They are beneficial for applications such as environment monitoring and animal monitoring.

This paper proposes a multi-tier framework for WSN-assisted IoT network deployment, together with the routing protocol. The primary objective of our work is to introduce a common framework for various types of WSN-assisted IoT applications. Standard work on static and dynamic protocols in the WSN-assisted IoT domain is included in the literature survey in section 2. The functionality of the proposed framework, communication constraints, and the various node scenarios are reported in section 3. The proposed routing protocols and all scenarios are described in section 4. The performance comparison of the proposed work with static and dynamic protocols for various metrics is covered in section 5. The paper concludes in section 6 with future scope.

9.2   RELATED WORK

WSN-assisted IoT networks are deployed on a large scale. Identifying and supervising large sensors across a vast network, as well as maintenance and servicing, are significant challenges. For example, a sensor can be attached to any dangerous species such as lions or rhinos, which can be continuously changing location, so in order to record animal activity, long-lasting attachable sensor batteries are needed. Energy-efficient communication is required to increase network lifetime [13, 14].

The sensor node's primary job is to send data to the base station via intermediate nodes. This task can be accomplished via direct routing, multi-hop routing and cluster routing. In direct routing, data can be sent directly from the sensor node to the base station (BS), but a long distance between sensor nodes and BS causes large energy consumption, the early death of sensor nodes and decreased network lifetime. In multi-hop routing, data is forwarded from sensor node to BS via intermediate sensor nodes. With multi-hop routing maintenance is greater but energy consumption less than in direct routing. Clustering routing divides the network into clusters, where some nodes such as cluster head (CH) and relay node (RN) are chosen for higher responsibility. Other nodes from the local cluster send their data to their CH, and the CH aggregates the data into a single packet and forwards it to the BS. The maintenance of clusters is not trivial and increases the additional overheads, particularly when the CH runs out of energy [15, 16].

Researchers have been busy modifying existing work to match the requirements of current techniques. For instance, the traditional WSN routing protocol is inefficient in the IoT network due to its vast size and scalability [12]. The LEACH protocol has been updated for better performance. In [17], a risk analysis is performed to decrease network overheads. The energy-efficient scheme proposed in [17] minimizes latency and balances security appropriately. TB-LEACH selects CHs based on time, organizing the clusters efficiently and increasing network lifetime by around 25% [18]. In [19], CHs are selected based on distance between nodes and base station, and lifetime is increased by 10%.

EA-CRP [20] is a multi-layer protocol that uses a division algorithm to enlarge the network and support short-distance communication. The proposed protocol maintains the cluster with reduced communication overheads. The CHs are selected based on the node residual energy ratio and the average distance between nodes. The SEEP protocol proposed in [21] is for static and dynamic application of WSN-assisted IoT large-scale networks. The performance of SEEP is slightly better than that of EA-CRP. In [22], the BPA-CRP protocol is proposed, which has four communication modes corresponding to different types of nodes in the network. In BPA-CRP CHs are selected by 'round robin', and all the normal nodes in the cluster take on this role one by one in their own time slot. BPA-CRP reduces the communication overheads for the cluster as there is less CH reselection. In [23], EA-DB-CRP is proposed for data gathering in WSN applications. The same node can perform the role of CH and RN but not in the same round.

PEGASIS is an improved version of LEACH. The nodes need to be arranged in the form of a chain in PEGASIS protocol, performed by base station or nodes. The implementation of PEGASIS is challenging, as knowledge of the entire network is needed [24]. HEED is a residual energy-based CH selection protocol, and its algorithm is based on distributive clustering [25]. Jafri et al. [26] proposed the MIEEPB protocol based on the sink mobility concept. The sink is moved to a predefined location (called sojourn location) for data collection from the chain leader and stays there for a predefined time (sojourn time). In MIEEPB, node chains can have multiple leaders. MIEEPB performs better than PEGASIS.

9.3 SYSTEM MODEL

9.3.1 PROPOSED FRAMEWORK

An example of the multi-layer hierarchical framework is given in Figure 9.1, which illustrates a smart health monitoring system in which a person wearing medical sensor devices is able to walk across a whole terrain. If their health-related information matches the criteria for critical condition, an alert message will be forwarded to the minimum-distance relay node (RN). The RN forwards the patient data to the local layer head (LH). The LH transmits the data to the upper layer coordinator (LCO). This process is repeated until the data reached the base station (BS). The BS sends the data and its location to the nearest hospital, which can arrange emergency medical assistance and equipment.

The scope of the proposed framework is not limited to smart healthcare monitoring systems. It can also be used for other WSN-assisted IoT applications, such as animal monitoring and environmental monitoring. A detailed description is given in the next section. This framework was proposed in [9, 14] for static applications. However, after analyzing the various applications, we used the mobility model (random walk and random waypoint) at node level in our simulation work in order to make the multi-layer framework a realistic dynamic application.

9.3.2 COMMUNICATION CONSTRAINTS

The communicating nodes in the multi-tier framework are as follows [14].

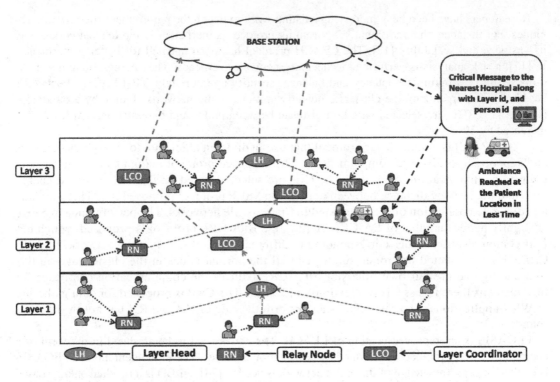

FIGURE 9.1 A multi-layer hierarchical framework.

9.3.2.1 Communication Constraints for Local Cluster

A node will not communicate within the local cluster under the following conditions:

1. if $x \in NNset_1$, $y \in NNset_1$ && Distance $(x,y) \leq r$
2. if $x \in NNset_1$, $y \in LHset_1$ && Distance $(x,y) \leq r$
3. if $x \in RNset_1$, $y \in RNset_1$ && Distance $(x,y) \leq R$

Communication can be established with the following conditions:

4. if $x \in NNset_1$, $y \in RNset_1$ && Distance $(x,y) \leq r$
5. if $x \in LHset_1$, $y \in RNset_1$ && Distance $(x,y) \leq r$

(NNs can communicate to LH via RNs).

9.3.2.2 Communication Constraints across Clusters

Nodes cannot communicate between one cluster and another under the following conditions:

1. if $x \in NNset_1$ in $LAYER_{Lower}$, $y \in NNset_2$ in $LAYER_{Upper}$ && Distance $(x,y) \leq r$
2. if $x \in NNset_1$ in $LAYER_{Lower}$, $y \in RNset_2$ in $LAYER_{Upper}$ && Distance $(x,y) \leq r$
3. if $x \in RNset_1$ in $LAYER_{Lower}$, $y \in RNset_2$ in $LAYER_{Upper}$ && Distance $(x,y) \leq R$
4. if $x \in NNset_1$ in $LAYER_{Lower}$, $y \in LHset_2$ in $LAYER_{Upper}$ && Distance $(x,y) \leq r$
5. if $x \in LHset_1$ in $LAYER_{Lower}$, $y \in LHset_2$ in $LAYER_{Upper}$ && Distance $(x,y) \leq r$
6. if $x \in RNset_1$ in $LAYER_{Lower}$, $y \in LHset_2$ in $LAYER_{Upper}$ && Distance $(x,y) \leq r$
7. if $x \in LHset_1$ in $LAYER_{Lower}$, $y \in LHset_2$ in $LAYER_{Upper}$ && Distance $(x,y) \leq r$

Communication is allowed under the following conditions:

1. if x ∈ LHset₁ in LAYER_Lower, y ∈ LCOset₂ in LAYER_Upper &&Distance (x,y) ≤ r, &&LAYER_id (LAYER_Lower) = (LAYER_id + 1)(LAYER_Upper)
2. if x ∈ LCOset₁ in LAYER_Lower, y ∈ LCOset₂ in LAYER_Upper&&Distance (x,y) ≤ r, &&LAYER_id (LAYER_Lower) = (LAYER_id + 1)(LAYER_Upper)
3. if x ∈ LCOset₁ in LAYER_Lower, y ∈ BS in LAYER_Upper&&Distance (x,y) ≤ r, &&LAYER_id (LAYER_Lower) = (LAYER_id + 1) (LAYER_Upper)
4. if x ∈ LHset₁ in LAYER_Lower, y ∈ BS in LAYER_Upper&&Distance (x,y) ≤ r, &&LAYER_id (LAYER_Lower) = (LAYER_id + 1) (LAYER_Upper)

9.3.3 Different Network Scenarios

The scope of the proposed multi-layer hierarchical framework is not limited to static WSN-assisted IoT applications as in [9, 14]. In real life the IoT can cover numerous applications with diverse node and network arrangement. We now extend the work of [9, 14] to consider three different network contexts.

Context 1. The participation nodes in this application are static after deployment. The algorithm of this context is given in algo. 1. Generally, this kind of network scenario is used in environment monitoring [9].

Context 2. The node's location remains dynamic in this scenario. The movement of the nodes is limited to the layer boundary. The graphical representation and algorithm are included in Figure 9.2 and algo. 2. This kind of network system is used in animal monitoring in zoos, and we are assuming an animal with a wearable sensor device. Different species of animals are looked after in their own areas [27].

Context 3. The nodes in this network context are free to move without restrictions across the entire network area. This network scenario is for a smart health monitoring system where a person is equipped with medical sensors [28]. The network scenario is illustrated in Figure 9.3, and the algorithm is given in algo. 3.

9.3.4 Assumptions

a. Nodes are randomly distributed in the network, and the network area is divided into equal-sized clusters. The nodes follow the static and dynamic scenarios discussed.
b. All the nodes have the same capability and resources. The energy is depleted in data communication processes in accordance with the radio energy model.

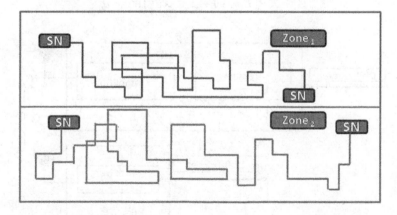

FIGURE 9.2   The movement of nodes within the local zone boundary (context 2).

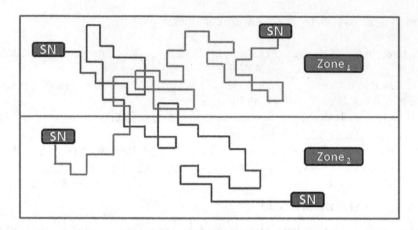

FIGURE 9.3 Node movement over the entire network (context 3).

   c. The sensing radius (R) of RN is slightly more than that of NN, LH, and LCO.

   d. Nodes are aware of their location in the network through the global positioning system (GPS) [9, 20].

   e. Nodes are equipped with a non-replaceable battery and have limited power energy only. The BS is equipped with sufficient resources [29].

   f. Nodes forward data in accordance with the communication constraints (see section 3.2).

   g. All nodes in the network are connected to each other and have a direct or multi-hop pathway to the BS.

9.3.5 ENERGY MODEL

Energy scavenging is the most significant aspect of the WSN-assisted IoT network, as sensor nodes run on limited power. Once the network is deployed, it remains unaltered, with data collection corresponding to various real-world situations. Sensor nodes have many functional parts including processor, battery, memory, transmitter, and transceiver. It is well known that of all these, the transmitter requires the greatest amount of energy. If the sensor node needs to send K bit data to another sensor node over distance d, energy consumption in the communication processes is computed by a one-order radio energy model (Figure 9.4) [9]. The equation for the transmitting and receiving process is given below.

$$E_{tx} = K \times \left(E_{elec} + E_{free\ space} \times d^2 \right) \text{if } d < d_0 \tag{9.1}$$

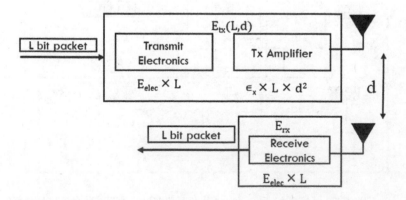

FIGURE 9.4 Radio energy model.

$$E_{tx} = K \times \left(E_{elec} + E_{multi\ path} \times d^4 \right) \text{if d} \geq d_0 \tag{9.2}$$

$$E_{rx} = K \times E_{elec} \tag{9.3}$$

Equation (9.1) and (9.2) represent energy consumption for short-distance and long-distance data transmission processes, respectively. E_{elec} represents transmitter and receiver circuitry. $E_{free\ space}$ and $E_{multi\ path}$ are node amplification in a free space and multi-path model, respectively. The communication threshold can be calculated as $\sqrt{\dfrac{E_{free\ space}}{E_{multi\ path}}}$. Energy depletion in data-receiving processes is identified in Equation (9.3).

9.3.6 NETWORK LIFETIME

We consider three matrices for network lifetime compared with standard routing protocols [30]:

- The total dead nodes after every round
- The round number when the first node becomes dead (FND_Statistics)
- The round number when the last node become dead, or the network becomes dead (LND_Statstics).

One round, when the BS receives the packet from NN, can be represented as (NN→RN→LH→LCO→BS).

ALGORITHM 1   ENERGY-EFFICIENT ROUTING (EER)

1: **Procedure** Selection of nodes for high responsibility and data transmission phase (total nodes N, layer head LH, layer coordinator LCO, total network layers M)
2: Random **N** nodes deployment in network
3: The network **divides** into **M** layers with equal size with fixed boundaries
4: **Declare** Layer_id for every layer   ▷ 1 to M Ids for every Layer
5: Layer Ids are assigned to every M layer, where Layer_id ≤ M.
6: **Declare** NN   ▷All NN ∈ N
7: **Declare** Local_layer   ▷Current selected layer
8: **declare** LHset = (initial empty)   ▷This set store LH of every layer
9: **declare** LCOset = (initial empty)   ▷ This set store LCO of every layer
10: User selects any Algo according to their application requirement from Algo 1 || Algo 2 || Algo3
11: **RN selection** for selected Algo (by ERNS method in [14])
12: LCO Selection Phase:
13:   **for** every layer, Layer_id ≤ M **do**
14:         Number of LCO in every layer =Layer_id-1
15:         **Elect** LCO from all nodes in local_layer (except layer 1) and selected LCO have the minimum distance to BS, LCO.energy ≥ energyThreshold
16:         LCO ∈ LCOset
17: **end for**

```
18: LH Selection Phase:
19:    for every layer, Layer_id ≤ M do
20:           LH Selection Phase:
21:           Elect one LH from all nodes in Local_layer and selected
              LH have high energy and the minimum distance to their
              LCO of LCOset in one above Upper_layer,
              LH.energy ≥ energyThreshold
22:           LH ∈ LHset
23:    end for
24: Routing Phase:
25:    for every layer, Layer_id ≤ M do
26:           transmit the data from a (a ∈ NN of Local_layer) to b
              (b ∈ RN of Local_layer)
27:           transmit the data from b → c → d → e (c ∈ LH, d ∈ LCO,
              e ∈ BS)
28:    end for
29:    for every layer, Layer_id ≤ M do
30:           if RN.energy ≤ energyThreshold then
31:            go to RN (ERNS in [14])
32:           end if
33:           if LH.energy and LCO.energy ≤ energyThreshold then
34:            go to line number 11 of algorithm 1
35:           else
36:            go to line number 21 of algorithm 1
37:           end if
38:    end for
39: end Procedure
```

9.4 ENERGY-EFFICIENT ROUTING

The scope of the proposed framework includes static and dynamic applications of WSN-assisted IoT, such as environment monitoring, animal monitoring and health monitoring. In this chapter it is assumed that nodes (animals or people with wearable sensors) will not leave the network area. The movement of nodes was discussed in section 3.3. In the simulation, a node's movement for both mobility models (random walk and random waypoint) is random in terms of movement direction and speed.

ALGORITHM 2   CONTEXT 1(STATIC NETWORK)

```
1: Nodes remain static in network
2:    for every layer, Layer_id ≤ M do
3:       for all NN ∈ Local_layer
4:          RN selection (by ERNS method in [14]) for Local_layer
5:       end for
6:    end for
```

Energy-efficient routing (EER) is given in algorithm 1. From line 1 to 10, nodes are deployed in the network and divided into equal-sized layer areas with an ID assigned to every layer. The user may choose any network context according to their application requirements. After the network type has been selected, relay nodes are selected for every layer by the effective relay node selection method (ERNS), based on the smallest distance from one node to other nodes in the layer [14]. The selected RN has R (sensing radius of RN) minimum distance from other nodes so that the layer can be covered with a smaller number of RNs. Node locations are updated for dynamic networks based on their functionality.

ALGORITHM 3   CONTEXT 2 (DYNAMIC NETWORK AND NODE MOVEMENT IS RESTRICTED TO THEIR LOCAL LAYER)

```
1: declare T = Traveling time of mobile node remains between 1 and t
2: declare θ = Traveling direction of mobile node remains between 0
   and 180
3: declare V = Traveling velocity of mobile node remains between 0
   and v
4: %%%% Node movement across their respective zone area %%%%%%%
5: Mobility phase:
6:   for every layer, Layer_id ≤ M do
7:   │   for all NN ∈ Local_layer
7:   │   │    P(x,y) ∈ NN
7:   │   │    xₙ = xₙ₋₁ + V× T × cosθn   ▷xₙ and xₙ₋₁new and previous x
   │   │                                  coordinate of node in Local_layer
8:   │   │    yₙ = yₙ₋₁ + V× T × sinθn   ▷yₙ and yₙ₋₁new and previous y
   │   │                                  coordinate of node in Local_layer
9:   │   │    NN ∈ P(xₙ,yₙ)
10:  │   end for
11:  end for
12: for every layer, Layer_id ≤ M do
13:  │   for all NN ∈ Local_layer
14:  │   │   RN selection (by ERNS method in [14]) for Local_layer
15:  │   end for
16: end for
```

ALGORITHM 4   CONTEXT3 (DYNAMIC NETWORK AND NODE MOVEMENT WITHIN ENTIRE NETWORK AREA)

```
1: declare T = Traveling time of mobile node remains between 1 and t
2: declare θ = Traveling direction of mobile node remains between 0
   and 180
3: declare V = Traveling velocity of mobile node remains between 0
   and v
4: %%%% Node movement across entire network area %%%%%%%
5: Mobility Phase:
6:  for all NN ∈N
7:  │      P(x,y) ∈ NN
7:  │    xₙ = xₙ₋₁ + V × T × cosθn   ▷xₙ and xₙ₋₁new and previous x
   │                                  coordinate of node in Local_layer
8:  │    yₙ = yₙ₋₁ + V × T × sinθn   ▷yₙ and yₙ₋₁new and previous y
   │                                  coordinate of node in Local_layer
9:  │    NN ∈ P(xₙ,yₙ)
10: end for
11: for every layer, Layer_id ≤ M do
12:  │   for all NN∈ Local_layer
13:  │   │  RN selection (by ERNS method in [14]) for Local_layer
14:  │   end for
15: end for
```

TABLE 9.1
Simulation Parameters

Parameters	Values
Total sensor nodes in network (N)	300
Network area	200×200 m^2
E_0 (nodes' initial energy)	0.5 J
E_{elec} for all the nodes	50 nj/bit
$E_{freespace}$ illustrates energy consumption for long-distance communication	0.0013 pJ/bit/m^4
$E_{multipath}$ illustrates energy consumption for short-distance communication	10 pJ/bit/m^2
Energy used in beam forming E_{bf}	5 nJ/bit
L (total bits in a single packet)	4000 bits
RN sensing radius	40 m
Number of layers	10
Speed interval	[0.2–2.2] m/s
Walk interval	[0–1] m/s
Pause interval	[4–6] s
Θ Direction interval	[−180, 180]

Furthermore, LCO and LC selections are made from line 11 to 23. The minimum distance node to BS and if node energy greater than the energy threshold will be selected as LCO. The number of LCOs in every layer is Layer_id-1. For instance, if the layer has id of five, then the number of LCOs is four. Later, LH is selected based on the smallest distance from a node to any LCO in the upper layer, and node energy needs to be more than the energy threshold. In the end, the packet is routed from NN to the nearest RN, RN to LH, and RN to LCO, and LCO to BS. The algorithm needs to be working even if some nodes are running below energy. The energy of the high-responsibility nodes (RN, LH, and LCO) is compared with the energy threshold in lines 29 to 38. If any of the nodes' energy is found to be less than the energy threshold, a new selection phase will be invoked. The algorithm will run until all the nodes consume their energy below the threshold (Table 9.1).

9.5   RESULT ANALYSIS AND DISCUSSION

In this work, the network is deployed with standard simulation parameters. The number of nodes in the network is 200, and the network is 200×200 m^2. The number of layers (M) in the network is 10. The layer area is 20×200 m^2. The nodes' mobility pattern is represented in Figures 9.5 and 9.6 via the random waypoint and random walk models. The initial position of the node is 1, and the final position after the simulation ends is 30. It can be observed that nodes travel greater distances via random walk than random waypoint. The reason is that the pause time always remains 0 in random walk. Hence, the movement of the node continues until the end of the simulation. But in the random waypoint case, the nodes remain constant in one position for a specific pause time (4–6 seconds in our simulation) then travel some distance at the selected velocity.

The network lifetime gains for static and dynamic nodes are illustrated in Figure 9.7. We can observe that a network with context 3 performs better than other scenarios. The network lifetime performance is also better when the node is dynamic by random waypoint. Context 2 performs the same as context 3 with both mobility models.

We also analyzed the network lifetime for static and dynamic context with FND and LND statistics (Figures 9.10 and 9.11). The last node in the network with context 1 expired at 1904 rounds, with context 2: RWP at 2109 rounds and context 2: RW after 1894 rounds. In context 3 with RWP and RW after 2599 and 2437 rounds, respectively. We can observe from Figure 9.9 that the network with context 1 has the first node expiring after 156 rounds, in context 2 with RWP and RW at 57 and

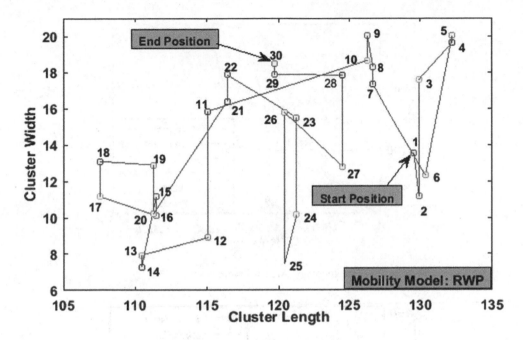

FIGURE 9.5   Mobility model: random waypoint.

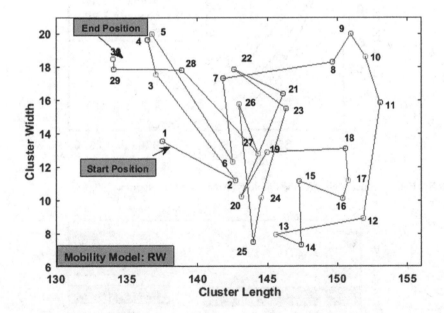

FIGURE 9.6   Mobility model: random walk.

79 rounds, and in context 3 with RWP and RW at 810 and 894 rounds, respectively. The results suggest that network lifetime with FND_Statistics under the random walk mobility model is better than under other models. But network lifetime is better for the LND metric with random waypoint mobility. The static topology (context 1) provides the shortest network lifetime, compared to the mobile topology.

After analyzing all the network context and mobility models, we found that context 3 performed better, and random waypoint enhances network lifetime compared to random walk for the following reasons. The nodes are static in context 1. Once the high-responsibility nodes are selected, data will

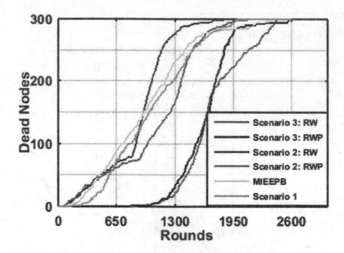

FIGURE 9.7   Network lifetime for the topology: Area = 400m², Nodes = 300.

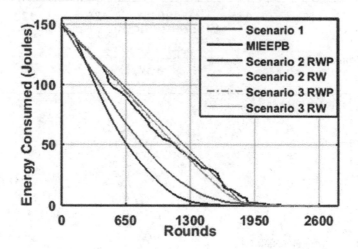

FIGURE 9.8   Energy consumption for the topology: Area = 400m², Nodes = 300.

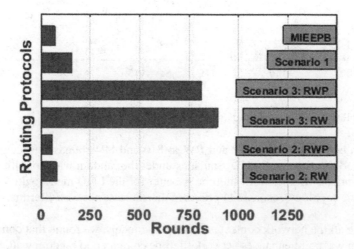

FIGURE 9.9   First node dead statistics for topology: Area = 400m², Nodes = 300.

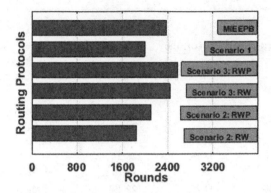

FIGURE 9.10   Last node dead statistics for topology: Area = 400m², Nodes = 300.

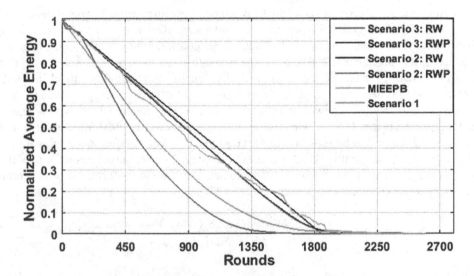

FIGURE 9.11   Comparison of normalized average energy consumption for topology: Area = 400m², Nodes = 300.

be forwarded continuously until the energy of one of the high-responsibility nodes falls below the threshold. In context 2, nodes change their location due to mobility, but the updated location remains in the same layer. Hence, high-responsibility nodes selected before mobility can provide coverage to all nodes in the local layers even after mobility due to mobility constraints. The new nodes will be selected as high-responsibility nodes until the energy of one of the nodes from RN, LH, and LCO drops below the threshold. In context 3, the nodes' position will be changed continuously after every round due to mobility. In the worst case, any layers in the network can be empty due to absence of mobility constraints. To provide node connectivity, RN, LH, and LCO will be selected after every round, making the energy depletion from the network uniform and parallel, and providing a better network lifetime.

We evaluated our result against the MIEEPB routing protocol proposed by Jafri [26]. MIEEPB also works on static topology, which is similar to our context 1. From Figures 9.7 and 9.10, it can be observed that MIEEPB provides a gain in network lifetime when compared to context 1. The last node expired in context 1 after 1994 rounds and in MIEEPB after 2377 rounds. The MIEEPB performance is impressive due to the mobility concept at the base station, which helps data collection from the nodes with minimum distance. Therefore, MIEEPB uses less energy and supports better network lifetime. But dynamic typology (context 2 and context 3) in a hierarchical framework increases

TABLE 9.2

Performance Gain of Proposed Protocols

Protocols	Context 1	Context 2 RWP	Context 2 RW	Context 3 RWP	Context 3 RW	MIEEPB
Scenario 1	-----	−5.7%	7.2%	−29.2%	−22.1%	−19.1%
Scenario 2: RWP	5.45%	----	12.3%	−22.9%	−15.5%	−12.7%
Scenario 2: RW	−7.8%	−14%	----	−39.3%	−31.8%	−28.5%
Scenario 3: RWP	22.6%	18.1%	28.2%	----	5.4%	7.7%
Scenario 3: RW	18.1%	13.4%	24.1%	−5.7%	----	2.4%
MIEEPB	16.1%	11.2%	22.2%	−8.4%	−2.5%	----

network performance over context 1and MIEEPB protocol; contexts 2 and 3 support better network lifetimes than MIEEPB. Figure 9.8 represents the total energy depletion of the compared protocols from first round to last round. The plot shows that energy consumption is almost equal in context 3: RWP, context 3: RW and context 2: RW. From the simulation results, it is observed that the residual energy of the network per round decreases gradually, but for each mobility model in context 3, energy consumption is better than in context 1, context 2 and MIEEPB. Energy depletion is uniform for all nodes in context 3 due to new node selection as RN in every round, which makes the network lifetime for context 3 better. The performance summary for all contexts with both mobility models is listed in Table 9.2, computed by Equation (9.4). This table compares performance gains for the different sets of deployment topology used in context 1, context 2 and context 3, along with MIEEPB. We can see from Table 9.2 that context 3: RWP continues to gain in performance terms compared with other scenarios (all values in positive).

$$\text{Performance gain} = \frac{LND\_Statistics_{protcol1} - LND\_Statistics_{protcol2}}{LND\_Statistics_{protcol2}} \times 100\% \qquad (9.4)$$

Figure 9.11 compares normalized average energy consumption of the static and mobile topology with MIEEPB protocol. The MIEEPB depletes less energy than context 1 and performs better. But for both mobility models, context 2 and context 3 show less energy consumption than context 1 and MIEEPB, and exhibit better network lifetime performance.

9.6 CONCLUSION

This chapter proposes a multi-tier framework with node mobility to enhance network lifetime. Realistic node deployment schemes are illustrated by context 1, context 2 and context 3, and a mobility model. From our analysis we conclude that context 3 is a better option, together with the random waypoint mobility model, in terms of network lifetime. We also propose an energy-efficient algorithm for the proposed framework to send data from the sensor nodes to base station. We believe that the proposed framework can meet the requirements of WSN-assisted IoT applications such as smart healthcare monitoring systems, and can also support core WSN-assisted IoT features such as scalability due to hierarchical topology. Base-station mobility has major benefits in increasing the network lifetime as compared to static sink.

Future work will seek to develop an improved base-station mobility model for data collection from the sensor nodes that could reduce communication distance and support enhanced network lifetime.

REFERENCES

1. Chase, J. (2013). The evolution of the internet of things. *Texas Instruments*, 1, 1–7.
2. Bandyopadhyay, D., & Sen, J. (2011). Internet of things: Applications and challenges in technology and standardization. *Wireless Personal Communications*, 58(1), 49–69.
3. Liu, G., & Jiang, D. (2016). 5G: Vision and requirements for mobile communication system towards year 2020. *Chinese Journal of Engineering*, *2016*(2016), 8.
4. Tan, L., & Wang, N. (2010). Future internet: The internet of things. In *2010 3rd International Conference on Advanced Computer Theory and Engineering (ICACTE)*, vol. 5, pp. V5–376.
5. Shen, J., Wang, A., Wang, C., Hung, P. C., & Lai, C. F. (2017). An efficient centroidbased routing protocol for energy management in WSN-assisted IoT. *IEEE Access*, 5, 18469–18479.
6. Preeth, S. S. L., Dhanalakshmi, R., Kumar, R., & Shakeel, P. M. (2018). An adaptive fuzzy rule based energy efficient clustering and immune-inspired routing protocol for WSN-assisted IoT system. *Journal of Ambient Intelligence and Humanized Computing*, 1–13. Doi:10.1007/s12652-018-1154-z
7. Wang, Z., Qin, X., & Liu, B. (2018, April). An energy-efficient clustering routing algorithm for WSN-assisted IoT. In *2018 IEEE Wireless Communications and Networking Conference (WCNC)*, pp. 1–6. IEEE.
8. Souissi, I., Azzouna, N. B., & Said, L. B. (2019). A multi-level study of information trust models in WSN-assisted IoT. *Computer Networks*, 151, 12–30.
9. Rani, S., Talwar, R., Malhotra, J., Ahmed, S. H., Sarkar, M., & Song, H. (2015). A novel scheme for an energy efficient Internet of Things based on wireless sensor networks. *Sensors*, 15(11), 28603–28626.
10. Manrique, J. A., Rueda-Rueda, J. S., & Portocarrero, J. M. (2016, December). Contrasting internet of things and wireless sensor network from a conceptual overview. In *2016 IEEE International Conference on Internet of Things (iThings) and IEEE Green Computing and Communications (GreenCom) and IEEE Cyber, Physical and Social Computing (CPSCom) and IEEE Smart Data (SmartData)*, pp. 252–257. IEEE.
11. Morillo, P., Orduña, J. M., Fernández, M., & García-Pereira, I. (2018). Comparison of WSN and IoT approaches for a real-time monitoring system of meal distribution trolleys: A case study. *Future Generation Computer Systems*, 87, 242–250.
12. Rani, S., Talwar, R., Malhotra, J., Ahmed, S. H., Sarkar, M., & Song, H. (2015). A novel scheme for an energy efficient Internet of Things based on wireless sensor networks. *Sensors*, 15(11), 28603–28626.
13. Baker, S. B., Xiang, W., & Atkinson, I. (2017). Internet of things for smart healthcare: Technologies, challenges, and opportunities. *IEEE Access*, 5, 26521–26544.
14. Shukla, A., & Tripathi, S. (2020). An effective relay node selection technique for energy efficient WSN-assisted IoT. *Wireless Personal Communications*, 112, 2611–2641.
15. Sajwan, M., Gosain, D., & Sharma, A. K. (2018). Hybrid energy-efficient multi-path routing for wireless sensor networks. *Computers & Electrical Engineering*, 67, 96–113.
16. Sajwan, M., Gosain, D., & Sharma, A. K. (2019). CAMP: Cluster aided multi-path routing protocol for wireless sensor networks. *Wireless Networks*, 25(5), 2603–2620.
17. Duan, J., Gao, D., Yang, D., Foh, C. H., & Chen, H. H. (2014). An energy-aware trust derivation scheme with game theoretic approach in wireless sensor networks for IoT applications. *IEEE Internet of Things Journal*, 1(1), 58–69.
18. Junping, H., Yuhui, J., & Liang, D. (2008, July). A time-based cluster-head selection algorithm for LEACH. In *2008 IEEE Symposium on Computers and Communications*, pp. 1172–1176. IEEE.
19. Kang, S. H., & Nguyen, T. (2012). Distance based thresholds for cluster head selection in wireless sensor networks. *IEEE Communications Letters*, 16(9), 1396–1399.
20. Darabkh, K. A., Al-Maaitah, N. J., Jafar, I. F., & Ala'F, K. (2018). EA-CRP: A novel energy-aware clustering and routing protocol in wireless sensor networks. *Computers & Electrical Engineering*, 72, 702–718.
21. Shukla, A., & Tripathi, S. (2020). A multi-tier based clustering framework for scalable and energy efficient WSN-assisted IoT network. *Wireless Networks*, 26, 3471–3493.
22. Darabkh, K. A., El-Yabroudi, M. Z., & El-Mousa, A. H. (2019). BPA-CRP: A balanced power-aware clustering and routing protocol for wireless sensor networks. *Ad Hoc Networks*, 82, 155–171.

23. Darabkh, K. A., Odetallah, S. M., Al-Qudah, Z., Ala'F, K., & Shurman, M. M. (2019). Energy-aware and density-based clustering and relaying protocol (EA-DB-CRP) for gathering data in wireless sensor networks. *Applied Soft Computing*, 80, 154–166.

24. Lindsey, S., & Raghavendra, C. S. (2002, March). PEGASIS: Power-efficient gathering in sensor information systems. In *Proceedings, IEEE Aerospace Conference*, Vol. 3, pp. 3–3. IEEE.

25. Younis, O., & Fahmy, S. (2004). HEED: A hybrid, energy-efficient, distributed clustering approach for ad hoc sensor networks. *IEEE Transactions on Mobile Computing*, 3(4), 366–379.

26. Jafri, M. R., Javaid, N., Javaid, A., & Khan, Z. A. (2013). Maximizing the lifetime of multi-chain pegasis using sink mobility. arXiv preprint arXiv:1303.4347.

27. Nóbrega, L., Tavares, A., Cardoso, A., & Gonçalves, P. (2018, May). Animal monitoring based on IoT technologies. In *2018 IoT Vertical and Topical Summit on Agriculture-Tuscany (IOT Tuscany)*, pp. 1–5. IEEE.

28. Catarinucci, L., De Donno, D., Mainetti, L., Palano, L., Patrono, L., Stefanizzi, M. L., & Tarricone, L. (2015). An IoT-aware architecture for smart healthcare systems. *IEEE Internet of Things Journal*, 2(6), 515–526.

29. Shukla, A., & Tripathi, S. (2018). An optimal relay node selection technique to support green internet of things. *Journal of Intelligent & Fuzzy Systems*, 35(2), 1301–1314.

30. Shukla, A. S., & Tripathi, S. (2019). A matrix-based pair-wise key establishment for secure and energy efficient WSN-assisted IoT. *International Journal of Information Security and Privacy (IJISP)*, 13(3), 91–105.

10 Multi-User Detection in Uplink Grant-Free NOMA with Dynamic Random Access Using Sinusoidal Sequences

*Shah Mahdi Hasan, Kaushik Mahata, and
Md Mashud Hyder*
The University of Newcastle, Australia

CONTENTS

DOI: 10.1201/9781003045809-12

10.1 INTRODUCTION

10.1.1 BACKGROUND AND MOTIVATION

The rapid growth in demand for wireless networks is no longer limited to the notion of enhanced throughput and privacy protection. Rather, it incorporates many verticals to cater for a wide range of distinct use cases. Massive machine-type communication (mMTC) is one of the major service categories of 5G, together with enhanced mobile broadband (eMBB), and ultra-reliable and low latency communications (URLLC), in the IMT 2020 Vision Framework of the International Telecommunication Union (ITU) [1]. This makes mMTC an extended use case of the 5G wireless network. A typical example of mMTC is the collection of a massive number of nodes, such as smart meters in the smart grid, IoT networks or automated factories [2].

Unlike human-agent-based wireless networks, mMTC systems are characterized by distinct features in terms of operation and traffic models: the network is dominated by uplink-bound traffic; and the transmitting nodes activate sporadically with the transmission of short-burst packets and low rates of data [2]. Moreover, the nodes (or agents) in an mMTC are usually equipped with low computational capacity and backup power. Grant-access-based legacy communication systems thus become irrelevant in the context of mMTC. For example, it has been shown in [3] that in 4G LTE, to transmit a small amount of data (e.g., 20 bytes of data using quadrature phase shift keying (QPSK) occupying 160 resource elements (REs)), the UL grant signaling overhead (i.e., 72 REs) can use around 30% of the total resources occupied by the UL grant and UL payload (i.e., 72 + 160 REs). Furthermore, these overheads impose higher latency, a further obstacle to sporadic transmission of short packets. To tackle these hindrances while supporting massive connectivity, UL grant-free non-orthogonal multiple access (NOMA) schemes, which have attracted significant attention among researchers in recent years, offer promising solutions [3–6]. In a grant-free NOMA system, resources are allocated among devices/users in a non-orthogonal manner. Furthermore, devices/ users transmit their data symbols in uplink without going through any complex hand-shaking procedure. This reduces transmission latency by lowering signaling overhead, one of the key features of the next-generation IoT.

Empirical findings show that even at peak hours of operation, the number of simultaneously transmitting nodes in an mMTC system is below 10% of all the configured users [7]. Thus, sparsity of user activation exists naturally in an mMTC system. The multi-user detection (MUD) problem in an mMTC paradigm is a sparse recovery problem under a compressive sensing (CS) framework. Most of the relevant literature considers a scenario where users simultaneously remain active or inactive across the entire random-access (RA) opportunity in a synchronous manner. This scenario was termed "framewise sparsity" in [8]. Based on this scenario, there has been significant development in low-complexity yet high-performance CS-MUD algorithms for use in mMTC networks. We argue that, in an uplink grant-free NOMA system, users should be able to enter and exit an RA opportunity freely in accordance with its requirement/design specifications. Not only does this reflect a more realistic use case in IoT, but it also alleviates the requirement to act synchronously during an RA opportunity. This chapter addresses this issue and develops a low-complexity yet high-performance multi-user detection framework to enable a grant-free uplink NOMA system.

10.1.2 RELATED WORKS

The brief literature survey in this section summarizes existing contributions and the overall direction of research in this area. The survey is sorted in chronological order so as to focus on the evolution of MUD algorithms over time.

The sparse nature of user activity in uplink mMTC systems means that only a small group of users among a massive number are attempting to transmit data in a random-access (RA) opportunity. This has led researchers to employ CS-based multi-user detection (CS-MUD) techniques for active user detection (AUD), channel estimation (CE), and data detection (DD) in mMTC systems. CS-MUD algorithms can be categorized into two subclasses: convex optimization-based sparse recovery; and iterative greedy algorithms. Due to their reduced computational effort and competitive performance, variants of greedy algorithms have played a major role in CS-MUD algorithm development.

One of the first examples of CS-MUD algorithms can be traced back to the work of Shim et. al. [9], who proposed a switching framework between greedy orthogonal matching pursuit (OMP) and linear minimum mean-square (LMMSE) detectors to obtain performance gain in a downlink code domain multiple access (CDMA) system. Most of the authors incorporated 'framewise sparsity' in the system model, where user activity remains unchanged over consecutive timeslots in an entire data transmission frame. This phenomenon was first exploited in [8], where the authors considered the MUD problem as a joint sparse multiple measurement vector (MMV) problem and developed low-complexity iterative order recursive least-squares (IORLS) algorithms for multi-user detection. In line with this, the authors of [10] suggested structured iterative support detection (SISD) for active user detection and data detection. The authors of [11] were the first to consider a more practical scenario by allowing users to enter the RA opportunity randomly. The temporal correlation of user activity over the frame was still considered as developing a dynamic CS-MUD algorithm, although the algorithm did not take account of users exiting the RA opportunity. The authors of [12] considered a framewise sparse system model to develop an approximate message passing (AMP) and expectation maximization (EM)-based MUD algorithm (AMP-EM), which demonstrated significant performance gain over SISD-based MUD. The authors of [13] considered the problem of users randomly entering and exiting an RA opportunity and argued that dynamic CS-MUD [11] blindly exploits the support estimate in previous timeslots. They developed an algorithm called prior information-aided adaptive subspace pursuit (PIA-ASP), where the common support information between two consecutive timeslots was used to enhance MUD performance. In [14], authors working in a similar direction proposed an adaptive direction method of multipliers (ADMM) for activity detection and data detection. In [15] the authors transformed the MMV problem into a block sparse single measurement vector (BS-SMV) problem and deployed a block sparsity adaptive subspace pursuit (BSASP) algorithm with a modified stopping criterion. In this work, as part of MUD, channel estimation of active users was also carried out. Based on this block-sparse signal model, two enhancements of BSASP were derived in [16], called threshold-aided BSASP (TA-BSASP) and cross-validation-aided BSASP (CVA-BSASP). These algorithms focused on reducing the frequency of misdetection and false alarm in activity detection by preventing the early/late termination of the greedy sparse recovery algorithm. The authors of [17] proposed a block sparse Bayesian learning (BSBL)-based MUD, where activity detection, channel estimation and data detection were carried out in a framewise sparse system model. In addition to the popular consideration of framewise sparsity, the authors of [18] incorporated the ternary nature of binary phase-shift keying modulation into their MUD algorithm, which was termed information-enhanced adaptive matching pursuit (IEAMP). A sparsity-blind greedy recovery algorithm called mSOMP-EXT was developed in [19], and did not require explicit knowledge of sparsity level (exact number of active users in a given RA opportunity) and noise statistics. The authors of [20] went down the Bayesian sparse recovery route by proposing expectation propagation (EP)-based joint user-activity detection and channel estimation. By approximating the computationally intractable Bernoulli–Gaussian distribution of user activity by a tractable multivariate Gaussian distribution, this finds an estimated *posteriori* distribution of the sparse channel vector, identifying the set of active users in the process. In a very recent work, the authors of [21] demonstrated the efficacy of long short-term memory (LSTM), a variant of deep learning architecture, for predicting user activities in an mMTC system. The activity prediction by LSTM was used as prior support information for initializing an OMP-based MUD algorithm. In this paper we refer to this MUD algorithm as LSTM-OMP.

TABLE 10.1
Summary of Existing MUD Algorithms

MUD algorithm	Framewise Sparsity	Channel Estimation	Required Priors
IORLS [8]	✓	✗	Channel gains, noise variance, sparsity level
SISD [10]	✓	✗	Channel gains
Dynamic CS-MUD [11]	✗	✗	Channel gains, sparsity level
AMP-EM [12]	✓	✗	Channel gains
PIA-ASP [13]	✗	✗	Channel gains
ADMM [14]	✗	✗	Channel gains
BSASP [15]	✓	✓	Noise statistics
TABASP [16]	✓	✗	Channel gains
BSBL [17]	✓	✓	None
IEAMP [18]	✓	✗	Channel gains, sparsity level
mSOMP-EXT [19]	✓	✗	Channel gains
EP [20]	✓	✓	Noise variance, channel gain variance
LSTM-OMP [21]	✓	✗	Channel gains, noise statistic

Table 10.1 summarizes the multi-user detection methods developed in the works mentioned, together with their required prior dependencies.

As can be seen from Table 10.1, only a few algorithms carry out channel estimation as part of MUD. Although [11, 13, 14] considered a dynamic random-access scenario, they are not capable of estimating the channels of active users randomly entering and exiting the RA opportunity. On the other hand, MUD frameworks that are equipped with a channel estimation capability [15, 17, 20] only consider the framewise sparse system model. None of the works listed in Table 10.1 considered a dynamic random-access scenario where perfect knowledge of channel gain of the active users is *not* available at the base station (BS). It is therefore interesting to consider this more practical scenario and develop a high-performance yet low-complexity MUD for it.

10.1.3 CONTRIBUTION

An mMTC network is dominated by sporadic uplink transmission of small packets where the users generally have low computational and power resources. To optimize the usage of scarce resources, the devices (or users) switch between sleep mode and active mode for short periods of time, resulting in a sporadic traffic model [15]. While most of the works in Table 10.1 assumed the availability of perfect channel information of all users at the BS, they ignored the fact that due to prolonged inactivity of the devices in an mMTC system, the channel information would be outdated. In fact, none of the works listed in Table 10.1 considered both dynamic random access and channel estimation.

The key motivation behind the present work is to address this vacuum. To be specific, we propose a multi-user detection framework for uplink grant-free mMTC systems with dynamic random access which does not require prior knowledge/statistics about the sparsity levels, channel and noise statistics. Below is a brief summary of our contribution:

- In this work, we develop AUD and user equipment (UE) activity detection (UAD) frameworks using sinusoidal sequences. Specifically, taking the sparse user activity in a random-access (RA) opportunity into account, we explore the utility of sparse parameter estimation algorithms like sparse iterative covariance estimation (SPICE) [22] for AUD in an mMTC system by using sinusoidal spreading sequences. Note that SPICE does not require framewise sparsity in the system model. Using sinusoidal sequences as the spreading sequences introduces a Vandermonde [23] structure in the received signal model. The resultant covariance matrix has a Topelitz structure and can be parameterized by a single-column vector which can be computed using fast Fourier transform (FFT). This leads to a significant reduction in computational cost and makes SPICE a competitive candidate for AUD. We further exploit the Vandermonde structure to create a data matrix with a non-singular covariance matrix. This enables us to use eigen-decomposition-based subspace estimation algorithms like estimation of signal parameters using rotational invariance techniques (ESPRIT) [24] as well.

- We devise a binary hypothesis test for the UE activity detection (UED) problem, based on the log-likelihood ratio (LLR). We develop sufficient statistics to derive the decision rule using the Neyman–Pearson criterion [25], which maximizes the probability of a correct decision for a given probability of false alarm. We give the close-form solution to enable simple threshold-aided detection of activities in each timeslot for each UE. We also show how to refine the set of estimated active UEs spelled out by the SPICE/ESPRIT-based MUD to obtain further performance gain in AUD.

- In conjunction with PSK modulation, we coherently process all the signal received in active timeslots to achieve better channel estimation. This is a modification of the framewise sparse scenario proposed in [26]. An additional advantage of this technique is that it allows us to provide sufficient conditions to categorize reliable and unreliable recovery of the transmitted data symbol. This may assist upper layers to deploy optimal power control strategies for the connected devices. In short, UEs which do not qualify for reliable recovery can receive an ARQ-like signal in downlink so that at the next RA opportunity they can secure reliable transmission.

- Finally, we carry out elaborate numerical investigations, mainly in realistic non-line of sight (NLOS) scenarios as demonstrated in 3GPP (release 9) so as to validate the efficacy of the proposed framework. In these trials, the proposed SPICE-MUD delivers satisfactory performance.

The rest of the paper is organized as follows. In Section 10.3 we demonstrate the system model of an UL mMTC cell equipped with a grant-free transmission mechanism. In Section 10.4, we describe the signal model of an RA opportunity with sinusoidal sequences. In Section 10.4, we discuss a computationally efficient SPICE algorithm for AUD. Next, we briefly discuss ESPRIT-based MUD in Section 10.5. Section 10.6 demonstrates the development of statistical hypothesis testing for UED, and in Section 10.7 we devise the conditions for categorizing the estimated set of UEs into reliable and unreliable sets by coherent processing of estimated channels. Section 10.8 contains all the numerical analysis where performance and complexity analysis are carried out. We discuss future research directions and conclude in Section X.

10.1.4 Notation

In this work, $\mathbb{B}, \mathbb{R}$ and $\mathbb{C}$ denote the fields of binary, real and complex numbers, respectively. Moreover, we denote a scalar, a vector and a matrix by x, x, $\mathbf{X}$ respectively. The jth column of a matrix $\mathbf{X}$ is represented by x_j. To denote the element in, for example, the mth row and n^{th} column of a matrix $\mathbf{X}$, we use parenthesis-based notation, i.e., $\mathbf{X}(m,n)$. The ℓ_m norm of a vector is denoted by $\|\cdot\|_m$ operator. An $M \times M$ identity matrix is given by $\mathbf{I}_M$. To represent a circular Gaussian distribution with mean vector μ and covariance matrix $\mathbf{A}$, we use the notation $\mathcal{CN}(\mu, \mathbf{A})$. $|\cdot|$ denotes absolute value or cardinality of a set depending on the context. The elementwise Hadamard product operator is denoted using $\odot$. Finally, the probability, expectation and variance operators are denoted as $\Pr(\cdot)$, $\mathbb{E}[\cdot]$ and $\mathrm{Var}[\cdot]$ respectively.

10.2   SYSTEM MODEL

Let us consider a single-antenna base station (BS) placed at the center of an mMTC cell with N user equipment (UE) configured. To enable NOMA, the system allocates $M < N$ orthogonal resources. In an mMTC system the number of UEs N can be very large; however, only a small fraction of the total N UEs become sporadically active and transmit information [7]. This fraction is commonly referred to in the literature as the activation ratio/probability p_a. In a grant-free NOMA system each UE $n \in \{0, 1, \cdots, N-1\}$ is assigned a unique spreading sequence which is an M dimensional complex-valued vector ϕ_n.

A random-access (RA) opportunity includes J continuous timeslots. When the nth UE becomes active in an RA opportunity and attempts to transmit a channel-encoded data symbol $\beta_{n,j}$, it uses ϕ_n to spread and modulate across M orthogonal resources. Hence, the entire set of modulated orthogonal resources can be given by $\phi_n\beta_{n,j}$. Note that we set $\beta_{n,j} = 0 \ \forall j$ for an inactive UE n. Also note that, since we are considering a dynamic RA scenario where UEs can start and stop transmitting data in any slot $j \in \{1, 2, \cdots J\}$, the set of active UEs might change in each timeslot, as shown in Figure 10.1. In this work, we define the set of active UEs over an RA opportunity thus: if a UE stops transmitting data and exits before the RA opportunity ends, it will be considered an active UE for that RA opportunity. The number of consecutive slots across which a UE remains active is denoted the activation period t_a. As will be shown later, the proposed method does not require homogeneous t_a for all UEs.

The received signal in the jth timeslot over M resources is denoted as $\left\{\mathbf{y}_j\right\}_{j=1}^{J}$ where $\mathbf{y}_j$ is given as

$$\mathbf{y}_j = \sum_{n \in \mathcal{N}}\phi_n\beta_{n,j}h_n + \mathbf{w}_j, \quad j = 1, 2, \ldots, J. \tag{10.1}$$

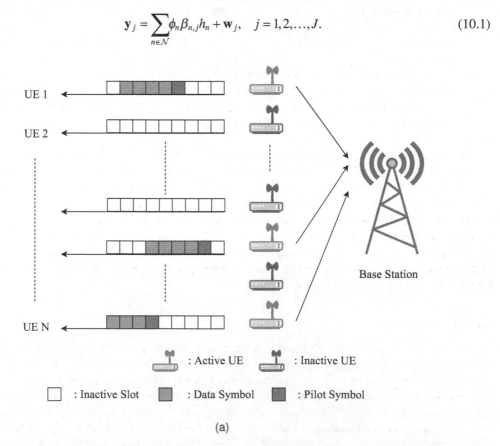

(a)

FIGURE 10.1   A dynamic random-access scenario where UEs enter and exit an RA opportunity freely with varying activation periods.

where $\mathcal{N}$ is the set of active UE indices during the RA opportunity. We are assuming flat Rayleigh channel fading gains where all M resources experience an equal, complex and scalar-valued channel gain h_n for the UE n. We also assume that the frame interval is much smaller than the channel coherence time. Hence the channels remain unchanged across the RA opportunity under consideration. Moreover, $\mathbf{w}_j$ denotes the complex vector-valued measurement noise. The measurement noise vectors $\{\mathbf{w}_n\}_{n=1}^{J}$ are modeled as zero-mean circularly Gaussian random vectors with the covariance matrix $\sigma^2\mathbf{I}$, where σ is not known to the BS and needs to be estimated.

Without any loss of generality, we can set the pilot symbol transmitted by an active UE $n \in \mathcal{N}$ as $\beta_{n,1} = 1$. Note that the pilot symbol is transmitted over the first timeslot of an activation period. However, for PSK constellations, $|\beta_{n,j}| = 1 \; \forall \; j$. It is important to note that the AUD step of the proposed framework is independent of the choice of constellation. However, as shown in later sections, choosing L-PSK modulation allows us to coherently process all the estimated channels to attain better data detection performance.

Next, we define the set of active UE indices $\mathcal{N}$ as the following

$$\mathcal{N} = \left\{b_1, b_2, \ldots, b_{|\mathcal{N}|}\right\}. \tag{10.2}$$

where the index of the nth active UE is denoted by b_n.

We can construct the following received signal matrix using $\{\mathbf{y}\}_{j=1}^{J}$

$$\mathbf{Y} = \left[\mathbf{y}_1 \; \mathbf{y}_2 \; \cdots \; \mathbf{y}_J\right].$$

where $\mathbf{Y}$ can be expressed as

$$\mathbf{Y} = \Phi\Upsilon + \mathbf{W}, \tag{10.3}$$

$\mathbf{W} = [\mathbf{w}_1 \; \mathbf{w}_2 \; \cdots \; \mathbf{w}_J]$ is the complex Gaussian noise matrix with rotational invariance. This particular property proves useful in creating signal snapshots, as we shall see. Moreover,

$$\Phi = \left[\phi_{a_1}, \; \phi_{a_2}, \; \cdots \; , \phi_{a_{|\mathcal{N}|}}\right], \tag{10.4}$$

Finally, $\Upsilon \in \mathbb{C}^{|\mathcal{N}| \times J}$ where

$$\Upsilon(n, j) = h_{b_n}\beta_{b_n,j}. \tag{10.5}$$

10.3   SYSTEM MODEL WITH SINUSOIDAL SEQUENCES

The use of sinusoidal spreading sequences in uplink grant-free NOMA systems was first proposed and demonstrated by the authors in [26], who showed that the MUD problem in an uplink grant-free NOMA system can be effectively formulated as a frequency estimation problem by using sinusoidal sequences. The ith element of a sinusoidal spreading vector ϕ_n has the following form

$$\phi_n(i) = \gamma \exp\left(i2\pi in/N\right) \; i \in \{0,1,\cdots M-1\}. \tag{10.6}$$

where γ is the transmit power. As will be shown later, the proposed framework can be extended to heterogeneous power allocation scenarios without any loss of generalization.

10.3.1 SIGNAL MODEL FOR SINUSOIDAL SPREADING SEQUENCES

Note that the angular frequency of the sinusoidal spreading sequence employed by UE n is given by

$$\omega_n = 2\pi n/N, \, n = 0, 1, \cdots, N-1. \tag{10.7}$$

With this, and (10.6), we can write the received signal at i^{th} resource of the jth frame as

$$\mathbf{y}_j(i) = \sum_{n \in \mathcal{N}} \exp(i\omega_n i) x_{n,j} + \mathbf{w}_j(i), \, i \in \{0,1,\cdots M-1\} \tag{10.8}$$

where, for brevity, we write

$$x_{n,j} = \gamma h_n \beta_{n,j} \tag{10.9}$$

Since, the number of timeslots J that form the RA opportunity is much smaller than the number of orthogonal resources M, i.e., $J \ll M$, the covariance matrix of the received signals $\{\mathbf{y}_j\}_{j=1}^{J}$ becomes singular. Subspace estimation techniques like ESPRIT rely on the eigen-decomposition of the covariance matrix of the received signals, hence it is not possible to detect UEs when $|\mathcal{N}| > J$, which is a common scenario in mMTC systems. However, it is important to note that the SPICE-MUD does not suffer from this concern since it does not rely on the eigen-decomposition of covariance matrices. Next, we describe how to create snapshots from the received signal matrix resulting in a non-singular covariance matrix. For further details, see [26, 27].

Let l be the length of each snapshot where $l > |\mathcal{N}|$. Although the proposed algorithm does not *a priori* require the number of active UEs $|\mathcal{N}|$, it requires the number of maximum possible active UEs in an RA opportunity to select an appropriate l. From each received signal vector $\mathbf{y}_j$ where $j \in \{1, 2 \cdots J\}$, we can express the resultant snapshot for any $i \in \{0, 1, \ldots M - l\}$ as:

$$\begin{aligned}
\varepsilon_{i,j} &\triangleq \left[\mathbf{y}_j(i) \; \mathbf{y}_j(i+1) \; \cdots \; \mathbf{y}_j(i+l-1) \right]^{\mathsf{T}} \\
&= \sum_{n \in \mathcal{N}} \varkappa_n x_{n,j} \exp(i\omega_n i) + \mathbf{w}_j(i:i+l-1).
\end{aligned} \tag{10.10}$$

In above the notation $\mathbf{w}_j(a:b)$ denotes a vector having a^{th} element of $\mathbf{w}_j$ as a starting component, all the way through to the bth element of $\mathbf{w}_j$ as the ending component. In addition, we define the following

$$\varkappa_n = \left[1 \; \exp(i\omega_n) \; \cdots \; \exp\{i(l-1)\omega_n\} \right]^{\mathsf{T}}. \tag{10.11}$$

From (10.10) we have the following

$$\Sigma_j \triangleq \left[\varepsilon_{0,j} \; \varepsilon_{1,j} \; \cdots \; \varepsilon_{M-l,j} \right] = \sum_{n \in \mathcal{N}} \varkappa_n \Theta_{n,j}^{\mathsf{T}} + \mathbb{W}_j, \tag{10.12}$$

where

$$\Theta_{n,j} \triangleq x_{n,j} \left[1 \; \exp(i\omega_n) \; \cdots \; \exp\{i(M-l)\omega_n\} \right]^{\mathsf{T}}, \tag{10.13}$$

$$\mathbb{W}_j = \begin{bmatrix} \mathbf{w}_j(0) & \mathbf{w}_j(1) & \cdots & \mathbf{w}_j(M-l) \\ \vdots & \vdots & \cdots & \vdots \\ \mathbf{w}_j(l-1) & \mathbf{w}_j(l) & \cdots & \mathbf{w}_j(M-1) \end{bmatrix}. \tag{10.14}$$

Let $\mathbb{K} \in \mathbb{B}^{l \times l}$ be an anti-diagonal matrix, i.e., the linear transformation $\mathbb{K}a$ flips the vector a upside down. Also, let $\mathrm{conj}(\cdot)$ denote the complex conjugate operator. Then we can write

$$\mathbb{K} \,\mathrm{conj}(\theta_n) = \theta_n \exp\{-i\omega_n(l-1)\}.$$

From (10.12) it yields

$$\mathbb{K} \,\mathrm{conj}(\Sigma_j) = \sum_{n \in \mathcal{N}} \varkappa_n \Theta^*_{n,j} \exp\{-i\omega_n(l-1)\} + \mathbb{K} \,\mathrm{conj}(\mathbb{W}_j),$$

where $(\cdot)^*$ denotes the conjugate transpose operator. Combining the above equation with (10.10), it yields

$$\bar{\Sigma}_j \triangleq \left[\Sigma_j \quad \mathbb{K} \,\mathrm{conj}(\Sigma_j)\right] = \sum_{n \in \mathcal{N}} \varkappa_n \bar{\Theta}^*_{n,j} + \bar{\mathbb{W}}_j, \tag{10.15}$$

where

$$\begin{aligned} \bar{\Theta}_{n,j} &= \left[\Theta^\mathsf{T}_{n,j} \quad \Theta^*_{n,j} e^{-i\omega_n(l-1)}\right]^*, \\ \bar{\mathbb{W}}_j &= \left[\mathbb{W}_j \quad \mathbb{K} \,\mathrm{conj}(\mathbb{W}_j)\right]. \end{aligned} \tag{10.16}$$

Combining the snapshot matrices $\left\{\bar{\Sigma}_j\right\}_{j=1}^J$ over the RA opportunity, we can write

$$\bar{\Sigma} = \left[\bar{\Sigma}_0 \quad \bar{\Sigma}_1 \cdots \bar{\Sigma}_{J-1}\right] = \sum_{n \in \mathcal{N}} \varkappa_n \chi^*_n + \bar{\mathbb{W}}, \tag{10.17}$$

where

$$\begin{aligned} \chi_n &= \left[\bar{\Theta}^*_{n,1} \quad \bar{\Theta}^*_{n,2} \quad \cdots \quad \bar{\Theta}^*_{n,J}\right]^*, \\ \bar{\mathbb{W}} &= \left[\bar{\mathbb{W}}_1 \quad \bar{\mathbb{W}}_2 \quad \cdots \quad \bar{\mathbb{W}}_J\right]. \end{aligned} \tag{10.18}$$

Having received the signals at the BS with a pre-specified choice of l, we can form $\bar{\Sigma}$ the Equations (10.10), (10.12), (10.15) and (10.17). Clearly, $\bar{\Sigma} \in \mathbb{C}^{l \times \mathcal{P}}$ where $\mathcal{P} = 2J(M+1-l)$.

10.3.2 SPARSE SIGNAL REPRESENTATION WITH SINUSOIDAL SEQUENCE

The signal received at the jth slot is written as

$$\mathbf{y}_j = \sum_{n=0}^{N-1} \phi_n x_{n,j} + \mathbf{w}_j, \tag{10.19}$$

where

$$\begin{cases} x_{n,j} = \gamma h_n \beta_{n,j}, & \text{if } n \in \mathcal{N} \\ x_{n,j} = 0, & \text{if } n \notin \mathcal{N} \end{cases}$$

As mentioned in Section 10.1.1, only a few of the total configured UEs are active in an RA opportunity, i.e., $|\mathcal{N}| \ll N$. Given the received signal at the BS with a suitable integer l, it is straightforward to construct the sparse representation of the data matrix $\overline{\Sigma}$ using the first equalities in Equations (10.10), (10.12), (10.15) and (10.17). Denoting this representation as $\overline{\mathbf{T}}$, we can write the following

$$\overline{\mathbf{T}} \triangleq \begin{bmatrix} \overline{\mathbf{T}}_0 & \overline{\mathbf{T}}_1 & \cdots & \overline{\mathbf{T}}_{J-1} \end{bmatrix} = \Omega F + \overline{\mathbb{W}} \tag{10.20}$$

where

$$\begin{aligned} \Omega &= \begin{bmatrix} \varkappa_0 & \varkappa_1 & \cdots & \varkappa_{N-1} \end{bmatrix}, \\ F &= \begin{bmatrix} \chi_0 & \chi_1 & \cdots & \chi_{N-1} \end{bmatrix}. \end{aligned} \tag{10.21}$$

In the above equation, F is a $N \times \mathcal{P}$ row-sparse matrix in which the all-zero rows belong to the inactive users in the RA opportunity under consideration.

10.4 SPICE-BASED AUD

10.4.1 FINDING ACTIVE USER INDICES USING SPICE

Sparse iterative covariance estimation (SPICE) was introduced by Stoica et al. [22] for sparse signal recovery, which is derived from a robust covariance fitting criterion. Since we have considered spatially and temporally uncorrelated and uniform measurement noise specified by the covariance matrix $\sigma^2 \mathbf{I}$, we use a specific modification of SPICE, termed SPICE+ in [22]. The covariance matrix of the snapshot vectors contained in data matrix $\overline{\mathbf{T}}$ can be written as

$$\mathbf{R} := \Omega \mathbb{P} \Omega^* + \sigma^2 \mathbf{I}, \tag{10.22}$$

where $\mathbf{P} = \text{diag}(p_0, p_1, \ldots, p_{N-1})$, where p_n denotes the associated unknown signal power of ω_n. SPICE+ considers the following covariance fitting criterion for the estimation of ω_n (see [28] and the reference therein) by updating the signal powers $\{p_0, p_1, \ldots p_{N-1}\}$ iteratively:

$$f = \left\| \mathbf{R}^{-\frac{1}{2}} \left(\hat{\mathbf{R}} - \mathbf{R} \right) \hat{\mathbf{R}}^{-\frac{1}{2}} \right\|_F^2, \tag{10.23}$$

where $\hat{\mathbf{R}}$ is the sample covariance matrix which is defined as

$$\hat{\mathbf{R}} = \frac{1}{\mathcal{P}} \overline{\Sigma \Sigma}^*. \tag{10.24}$$

Below we summarize the steps that take place in each iteration of SPICE+ for active user detection. The superscripts indicate iterative indices of the algorithm. The algorithm can be initialized using the power estimates $\left\{p_n^{(0)}\right\}_{n=0}^{N-1}$ obtained via periodogram [27] which can be computed using N-point FFT. It was shown in [22] that a reasonable initial estimate of noise variance can be computed as

$$\sigma^{2(0)} = \sum_{n=1}^{l} \frac{\tilde{p}_n^{(0)} \|\tilde{\varkappa}_n\|^2}{l}, \tag{10.25}$$

where the set $\left\{\tilde{p}_n^{(0)}\right\}_{n=1}^{l}$ and $\{\tilde{\varkappa}_n\}_{n=1}^{l}$ denotes the smallest l values of $\left\{p_n^{(0)}\right\}_{n=0}^{N-1}$ and corresponding truncated spreading sequences respectively. The following steps take place sequentially at the ith iteration:

1. Compute $\mathbb{P}^{(i-1)}$ using the signal power estimates $\left\{p_n^{(i-1)}\right\}_{n=0}^{N-1}$ estimated in the $(i-1)^{th}$ iteration. Update $\mathbf{R}^{(i)}$ using $\mathbb{P}^{(i-1)}$ and noise variance $\sigma^{2(i-1)}$ using (10.22):

$$\mathbf{R}^{(i)} = \Omega \mathbb{P}^{(i-1)} \Omega^* + \sigma^{2(i-1)} \mathbf{I}. \tag{10.26}$$

2. Using the most recent estimation $\mathbf{R}^{(i)}$ calculated in step 1 along with the signal power estimates $\left\{p_n^{(i-1)}\right\}_{n=0}^{N-1}$ and noise variance estimate $\sigma^{2(i-1)}$, compute the auxiliary variable $\rho^{(i)}$ as following

$$\rho^{(i)} = \sum_{n=0}^{N-1} \delta_n^{1/2} p_n^{(i-1)} \left\| \varkappa_n^* \mathbf{R}^{(i)-1} \hat{\mathbf{R}}^{1/2} \right\|_F + \tau^{1/2} \sigma^{2(i-1)} \left\| \mathbf{R}^{(i)-1} \hat{\mathbf{R}}^{1/2} \right\|_F, \tag{10.27}$$

where

$$\delta_n := \varkappa_n^* \hat{\mathbf{R}}^{-1} \varkappa_n / \mathcal{P}, \tag{10.28}$$
$$\tau := \mathrm{Tr}\left(\hat{\mathbf{R}}^{-1}\right) / \mathcal{P}.$$

Note that $\left\{\delta_n\right\}_{n=0}^{N-1}$ and τ remains constant throughout the iterations.

3. Using the auxiliary variable $\rho^{(i)}$, first update the estimate of noise variance

$$\sigma^{2(i)} = \sigma^{2(i-1)} \frac{\left\| \mathbf{R}^{(i)-1} \hat{\mathbf{R}}^{1/2} \right\|_F}{\tau^{1/2} \rho^{(i)}}. \tag{10.29}$$

4. Finally, update the signal power estimate associated with each ω_n for $n = 0, 1, \ldots, N-1$ using $\rho^{(i)}$ and $\sigma^{2(i)}$ as following

$$p_n^{(i)} = p_n^{(i-1)} \frac{\left\| \varkappa_n^* \mathbf{R}^{(i)-1} \hat{\mathbf{R}}^{1/2} \right\|_F}{\delta_n^{1/2} \rho^{(i)}}. \tag{10.30}$$

5. Continue steps 1–4 until the following criterion is satisfied: $\|\mathbf{p}^{(i)} - \mathbf{p}^{(i-1)}\| / \|\mathbf{p}^{(i-1)}\| \geq \varepsilon$, where ε is predefined tolerance.

An experimentally selected threshold α that minimizes the activity error rate (AER) (defined in a later section) is chosen to select the peaks associated with active user indices from the spectrum estimated by SPICE+. In other words, the estimated active user set can be defined as

$$\hat{\mathcal{N}} = \left\{ n \mid \log\left(\| p_n \|\right) > \alpha \right\}.$$

10.4.2 FAST COMPUTATION OF R USING FFT

Significant reduction of computational overhead can be accomplished in SPICE by using FFT [29]. Let us denote $\bar{\mathbf{R}} := \Omega\mathbb{P}\Omega^*$. Since Ω has Vandermonde structure, $\bar{\mathbf{R}}$ is a Hermitian Toeplitz matrix which yields the Hermitian Toeplitz structure of $\mathbf{R}$. Consequently both these matrices can be fully described by their first column vector. Now we can write the following

$$\bar{\mathbf{R}} = \Omega\mathbb{P}\Omega^* = \sum_{n=0}^{N-1} p_n \varkappa_n \varkappa_n^*$$

$$= \begin{bmatrix} r_0 & r_1 & \cdots & r_{l-1} \\ r_1^* & r_0 & \cdots & r_{l-2} \\ \vdots & \vdots & \ddots & \vdots \\ r_{l-1}^* & r_{l-2}^* & \cdots & r_0 \end{bmatrix}. \tag{10.31}$$

Each element of the matrix in the second equality of (10.31) can be expressed as

$$r_i = \sum_{n=0}^{N-1} p_n \exp\left(\frac{i2\pi in}{N}\right), \quad \text{where } i = 0,1,\ldots,l-1. \tag{10.32}$$

As a consequence, $\{r_i\}_{i=0}^{l-1}$ can be obtained as the following: compute N-point FFT of $\{p_n\}_{n=0}^{N-1}$. Then retain complex conjugates of first l elements. Upon obtaining $\{r_i\}_{i=0}^{l-1}$, the first column vector of $\mathbf{R}$ as specified in (10.22) can be obtained by simply adding σ^2 with the first element. Thus, step (1) in the SPICE iteration can be computed in a computationally efficient manner.

10.5 SUBSPACE ESTIMATION-BASED FAST AUD

Before describing subspace estimation-based AUD, let us define the following lemma:

Lemma 1: The rank of the noise-free part of $\bar{\Sigma}$ cannot be more than $|\mathcal{N}|$ if $l > |\mathcal{N}|$.

Proof: See Appendix A.

This particular property is exploited by the family of subspace algorithms where the first step is to obtain a low-rank approximation of $\bar{\Sigma}$. However, subspace algorithms require the number of active UEs $|\mathcal{N}|$ to find the $|\mathcal{N}|$-rank approximation by singular value decomposition (SVD). It is by no means trivial to establish this information at the BS; however, the problem was formulated as a model-order detection problem in [30] and was solved by minimizing the information criteria [31, 32]. Below we present a brief analysis of this topic.

10.5.1 Estimating the Number of Active UEs $|\mathcal{N}|$

As shown in [26, Appendix B], $\text{rank}\left(\overline{\Sigma}-\overline{W}\right)=|\mathcal{N}|$. Given $\overline{\Sigma}$, we can formulate estimating $|\mathcal{N}|$ as a model-order selection problem. Specifically, we chose to employ an information-criteria-based model-order selection paradigm, specifically because of its use of SVD. The individual calculated values can be used later in the estimation of $\mathcal{N}$.

The information-criteria-aided model-order selection problem has the following form

$$|\hat{\mathcal{N}}| = \arg\min_{k} -\log f\left(\overline{\Sigma}, k\right)+\mathcal{W}_k. \tag{10.33}$$

The log-likelihood function $f\left(\overline{\Sigma},k\right)$ in (10.33) has the following closed form as derived in [30]

$$\log f\left(\overline{\Sigma},k\right)=(l-k)\mathcal{P}\ \ln\left\{\frac{\prod_{i=k+1}^{l}\left(\varsigma_i^2\right)^{\frac{1}{(l-k)}}}{\frac{1}{(l-k)}\sum_{i=k+1}^{l}\varsigma_i^2}\right\}, \tag{10.34}$$

where $\{\varsigma\}_{i=1}^{l}$ denotes the l non-zero singular values of the $l\times\mathcal{P}$ matrix $\overline{\Sigma}$ in descending order without any loss of generality.

$\mathcal{W}_k$ in (10.33) is known as the penalty/bias correction term. Depending on the choice of $\mathcal{W}_k$, there exists an array of information criteria for different scenarios [30, 33–35]. The most commonly used is the Akaike information criterion (AIC) [32], for which $\mathcal{W}_k=k\left(2l-k\right)$ [30]. One of the major concerns posed by AIC is its tendency to overestimate the model order. Overestimating the number of active UEs will eventually lead to false detection of UEs. Considering that perspective, we instead employ the Bayesian information criterion (BIC) [36] for which

$$\mathcal{W}_k=\frac{1}{2}k\left(2l-k\right)\log\mathcal{P}. \tag{10.35}$$

By calculating log likelihood and the penalty term using the closed forms provided in (10.34) and (10.35), respectively, it is possible to compute the loss function given in (10.33) for different candidate values of k. By selecting the k which minimizes (10.33), one can readily estimate the number of active UEs $|\hat{\mathcal{N}}|$.

10.5.2 Estimating the Active UE Indices

In the subspace-based AUD, finding the active UE indices is equivalent to solving the underlying frequency estimation problem. Many candidate subspace estimation algorithms are detailed in the literature, e.g., ESPRIT, MUSIC, root-MUSIC, and Capon [27]. Among those, we employ forward-backward ESPRIT [24]. It has been shown statistically that ESPRIT outperforms other candidate algorithms and it is consequently recommended as the algorithm of choice for frequency estimation problems [27]. The details of ESPRIT implementation are well understood (see [24, 26, 27, 37]).

Let $\hat{\omega}_n$ denote the angular frequency of nth estimated active UE. To find the active UE index $\hat{a}_n$ we use the following definition

$$\hat{a}_n = \text{round}\left(\frac{N\hat{\omega}_n}{2\pi}\right).$$

In other words, the estimated set of active UE $\hat{\mathcal{N}}$ can be written as

$$\hat{\mathcal{N}} = \left\{ \hat{b}_1, \hat{b}_2, \ldots, \hat{b}_{|\hat{\mathcal{N}}|} \right\}.$$

10.6   USER ACTIVITY DETECTION OVER RA OPPORTUNITY

In this section, we demonstrate the framework for User Activity Detection in a given RA opportunity. In what follows next, we will derive sufficient statistics to develop the decision rule about detecting UE activity using the Neyman–Pearson criterion [25].

10.6.1   STATISTICS SUFFICIENT FOR USER ACTIVITY DETECTION

Using the detected active UE indices, we construct the $M \times |\hat{\mathcal{N}}|$ matrix $\hat{\Phi}$ such that

$$\left[\hat{\Phi} \right](m,n) = \gamma \exp\left(i2\pi m\hat{b}_n / N \right). \tag{10.36}$$

Subsequently we calculate the linear least-squares estimate of Υ, see (10.3),

$$\hat{\Upsilon} = \left(\hat{\Phi}^* \hat{\Phi} \right)^{-1} \hat{\Phi}^* \mathbf{Y}. \tag{10.37}$$

where $\hat{\Upsilon} \in \mathbb{C}^{|\hat{\mathcal{N}}| \times J}$. Let us denote $\mathbf{O} = \left(\hat{\Phi}^* \hat{\Phi} \right)^{-1} \hat{\Phi}^* \mathbb{W}$ i.e., the noise term of (10.37). The nth row of $\mathbf{O}$ can be written as

$$\mathbf{o}_n = \mathbf{e}_n \left(\hat{\Phi}^* \hat{\Phi} \right)^{-1} \hat{\Phi}^* \mathbb{W}, \tag{10.38}$$

where $\mathbf{e}_n$ is the nth row of the identity matrix. Let us write $\mathbf{q}_n = \mathbf{e}_n \left(\hat{\Phi}^* \hat{\Phi} \right)^{-1} \hat{\Phi}^*$. Then, it follows

$$\mathbf{o}_n = \mathbf{q}_n \mathbb{W}. \tag{10.39}$$

We can transform the row vector $\mathbf{o}_n$ into a column vector by vectorizing it. This yields the following

$$\mathrm{vec}\left(\mathbf{o}_n \right) = \left(\mathbf{I} \otimes \mathbf{q}_n \right) \mathrm{vec}\left(\mathbb{W} \right) \tag{10.40}$$

where $\mathbf{I}$ is the identity matrix of appropriate size. Now let us calculate the covariance of $\mathrm{vec}(\mathbf{o}_n)$. It is straightforward to validate that $\mathrm{E}[\mathrm{vec}(\mathbf{o}_n)] = \mathbf{0}$. So the covariance matrix of $\mathrm{vec}(\mathbf{r}_n)$ can be computed as

$$\begin{aligned} \mathrm{cov}\left[\mathrm{vec}\left(\mathbf{o}_n \right) \right] &= \mathrm{E}\left[\mathrm{vec}\left(\mathbf{r}_n \right) \mathrm{vec}\left(\mathbf{r}_n \right)^* \right] \\ &= \left(\mathbf{I} \otimes \mathbf{q}_n \right) \mathrm{E}\left[\mathrm{vec}\left(\mathbb{W} \right) \mathrm{vec}\left(\mathbb{W} \right)^* \right] \left(\mathbf{I} \otimes \mathbf{q}_n \right)^*. \end{aligned} \tag{10.41}$$

Noting that $\mathrm{vec}\left(\mathbb{W}\right) = \left[\mathbf{w}_1^T \;\; \mathbf{w}_2^T \;\; \cdots \;\; \mathbf{w}_J^T\right]^T$, we can write the following

$$
\mathrm{E}\left[\mathrm{vec}\left(\mathbb{W}\right)\mathrm{vec}\left(\mathbb{W}\right)^*\right] = \begin{bmatrix} \mathrm{E}\left[\mathbf{w}_1\mathbf{w}_1^*\right] & \mathrm{E}\left[\mathbf{w}_1\mathbf{w}_2^*\right] & \cdots & \mathrm{E}\left[\mathbf{w}_1\mathbf{w}_J^*\right] \\ \mathrm{E}\left[\mathbf{w}_2\mathbf{w}_1^*\right] & \mathrm{E}\left[\mathbf{w}_2\mathbf{w}_2^*\right] & \cdots & \cdots \\ \vdots & \vdots & \ddots & \vdots \\ \mathrm{E}\left[\mathbf{w}_J\mathbf{w}_1^*\right] & \cdots & \cdots & \mathrm{E}\left[\mathbf{w}_J\mathbf{w}_J^*\right] \end{bmatrix}.
\tag{10.42}
$$

Since complex valued measurement noise vectors $\mathbf{w}_1, \mathbf{w}_2, \cdots, \mathbf{w}_j$ are assumed to be mutually independent, the following relationship holds

$$
\mathrm{E}\left[\mathbf{w}_c\mathbf{w}_d^*\right] = \begin{cases} \sigma^2\mathbf{I}_M & \text{if } c = d \\ \mathbf{0}_M & \text{if } c \neq d, \end{cases}
\tag{10.43}
$$

where $\mathbf{0}_M$ and $\mathbf{I}_M$ are $M \times M$ zero matrix and identity matrix respectively. This yields the following

$$
\mathrm{E}\left[\mathrm{vec}\left(\mathbb{W}\right)\mathrm{vec}\left(\mathbb{W}\right)^*\right] = \begin{bmatrix} \sigma^2\mathbf{I}_M & \mathbf{0}_M & \cdots & \mathbf{0}_M \\ \mathbf{0}_M & \sigma^2\mathbf{I}_M & \cdots & \cdots \\ \vdots & \vdots & \ddots & \vdots \\ \mathbf{0}_M & \cdots & \cdots & \sigma^2\mathbf{I}_M \end{bmatrix}
\tag{10.44}
$$

$$
= \sigma^2\mathbf{I}_{NJ},
$$

where $\mathbf{I}_{NJ}$ is an $NJ \times NJ$ identity matrix. Plugging this value in (10.41) followed by using the mixed-product property of Kronecker product, i.e., $(\mathbf{A} \otimes \mathbf{B})(\mathbf{C} \otimes \mathbf{D}) = (\mathbf{AC}) \otimes (\mathbf{BD})$, we can write the following

$$
\mathrm{cov}\left[\mathrm{vec}\left(\mathbf{r}_n\right)\right] = \sigma^2\mathbf{I}^2 \otimes \left(\mathbf{q}_n\mathbf{q}_n *\right)
\tag{10.45}
$$

$$
= \sigma^2\mathbf{q}_n\mathbf{q}_n^*\mathbf{I}.
$$

This implies that $\mathrm{vec}(\mathbf{r}_n)$ is a complex valued random vector with circularly symmetric normal elements each with mean $\mu = 0$ and variance $\mathbf{q}_n\mathbf{q}_n^*\sigma^2$. Note that σ^2 can be estimated from the trailing singular values ς_i (see Section 10.5). To be specific

$$
\hat{\sigma}^2 = \frac{1}{l - |\hat{\mathcal{N}}|} \sum_{i=|\hat{\mathcal{N}}|+1}^{l} \varsigma_i^2.
\tag{10.46}
$$

Moreover, SPICE estimates noise variance as a part of AUD, as shown in the previous section. With this development, we can devise a hypothesis test based on the likelihood ratio on each timeslot j of detecting the presence of an active UE in that timeslot. To be specific, a detected UE $n \in \hat{\mathcal{N}}$ does not transmit data in the jth timeslot under null hypothesis $\mathbf{H}_0$, whereas the alternative hypothesis $\mathbf{H}_1$ indicates that the UE n is indeed transmitting data in the jth timeslot. In this setting, the underlying models in each hypothesis are described by circular Gaussian density functions having different variances, indicating the absence and presence of a data symbol, respectively. We can write these hypotheses as

$$\mathcal{H}_0 : \hat{\Upsilon}(n,j) \sim \mathcal{CN}\left(0,\sigma_1^2\right)$$

$$\mathcal{H}_1 : \hat{\Upsilon}(n,j) \sim \mathcal{CN}\left(0,\sigma_2^2\right)$$

where $\sigma_1^2 = \hat{\sigma}^2 \mathbf{q}_n \mathbf{q}_n^*$ and $\sigma_1 < \sigma_2$. Since both models under these hypotheses are circular Gaussian density functions differing only by variance, it can be shown that the sufficient statistic $\Lambda_{n,j}$ for the likelihood ratio test can be written as

$$\Lambda_{n,j} = \left|\hat{\Upsilon}(n,j)\right|^2 \quad \text{where} \quad j \in \{1, 2, \ldots J\}, \tag{10.47}$$

which can be described using a scaled chi-square density function with 2 degrees of freedom accounting for the real and imaginary parts. This density can be expressed using Gamma distribution with shape-scale parameters. Hence, as shown below, the test can be formulated as

$$\mathcal{H}_0: \Lambda_{n,j} \sim \Gamma\left(1,\sqrt{2}\sigma_1\right)$$

$$\mathcal{H}_1: \Lambda_{n,j} \sim \Gamma\left(1,\sqrt{2}\sigma_2\right)$$

It should be noted that to carry out the null hypothesis test we do not need to know the statistics under $\mathcal{H}_1$, i.e., the channel statistics. We use the Neyman–Pearson criterion for devising a decision threshold for each active UE $n \in \hat{\mathcal{N}}$. The Neyman–Pearson criterion maximizes the probability of active UE detection for a pre-specified false alarm probability P_{FA}. According to the criterion, decision threshold α_n of UE n for a given false alarm probability P_{FA} can be obtained by solving the following [25]:

$$P_{FA} = \int_{\alpha_n}^{\infty} \Gamma\left(1,\sqrt{2}\sigma_1\right) d\Lambda$$
$$= \int_{\alpha_n}^{\infty} \frac{1}{\sqrt{2}\sigma_1} \exp\left(-\frac{\Lambda}{\sqrt{2}\sigma_1}\right) d\Lambda. \tag{10.48}$$

Now we can compute a binary activation matrix $\mathbf{A} \in \mathbb{B}^{|\mathcal{N}| \times J}$, which indicates the activity of an active user $n \in \hat{\mathcal{N}}$ in a timeslot $j \in \{0, 1, \ldots J - 1\}$. It can be defined as

$$\begin{cases} \mathbf{A}_{n,j} = 1, & \text{if } \Lambda_{n,j} > \alpha_n \\ \mathbf{A}_{n,j} = 0, & \text{if } \Lambda_{n,j} < \alpha_n \end{cases}$$

Finally, we update the least-squares estimate $\hat{\Upsilon}$ as the following:

$$\hat{\Upsilon} = \hat{\Upsilon} \odot \mathbf{A}. \tag{10.49}$$

where $\odot$ denotes the element-wise Hadamard product.

10.6.2 REFINING THE ACTIVE USER SET

While computing the least-squares estimate in (10.37), the received signal matrix $\mathbf{Y}$ was projected onto the column space of $\hat{\Phi}$. The presence of falsely detected users in the estimated active user set $\hat{\mathcal{N}}$ and the consequent inclusion of a corresponding sinusoidal sequence in $\hat{\Phi}$ leads to the following phenomenon. Since in this case the received signal does not have the signal components of falsely included active user indices, it only projects noise onto the subspace spanned by the spreading sequences associated with those falsely detected users. In this case, for the falsely detected users, the likelihood ratio test fails to reject the null hypothesis across all timeslots, i.e., the entire RA opportunity, and as a result, the activation matrix $\mathbf{A}$ contains corresponding all-zero rows. Based on this criterion, it is possible to further refine the estimated active user set $\hat{\mathcal{N}}$ spelled out by ESPRIT or SPICE in Sections 10.3 and 10.4. Hence, the refined estimated active user set can be defined as

$$\tilde{\mathcal{N}} = \left\{ n \in \hat{\mathcal{N}} \mid \left\| \mathbf{e}_n \mathbf{A} \right\|_0 > 0 \right\}. \tag{10.50}$$

In what follows, we construct $\tilde{\Phi}$ using $\tilde{\mathcal{N}}$ and revise the least-squares estimate of $\mathbf{Y}$,

$$\tilde{\mathbf{Y}} = \left(\tilde{\Phi}^* \tilde{\Phi} \right)^{-1} \tilde{\Phi}^* \mathbf{Y}. \tag{10.51}$$

We use $\tilde{b}_n$ to denote the nth estimated active user index. In other words,

$$\tilde{\mathcal{N}} = \left\{ \tilde{b}_1, \tilde{b}_2, \ldots, \tilde{b}_{|\tilde{\mathcal{N}}|} \right\}.$$

10.7 CHANNEL ESTIMATION WITH DYNAMIC RA

In this section, we describe channel estimation and data detection methods. Since one of the defining features of mMTC is short-packet transmission over an uplink wireless channel, low-order modulation schemes like L-PSK are the primary choices for signal constellation. It is possible to coherently process all signals over active timeslots for improved channel gain estimation. In this way, it is possible to mitigate the perturbation of noise in the channel gain estimates. It was shown in [26] that the coherent processing of channel estimates provides performance gains. For other modulation schemes like L-QAM, one can employ pilot-aided channel estimation (PACE) as described in [9, 20, 38].

10.7.1 CHANNEL ESTIMATION

Let us define the set of active timeslots $\mathcal{J}_n$ for $n \in \tilde{\mathcal{N}}$ where the first element denotes the pilot slot followed by data slots. For brevity, j will be used to index $\mathcal{J}_n$ without subscript in the following details, where $j = 1$ denotes the first appearance of a non-zero element in the timeslot for a UE n, i.e., the pilot slot. Let us define the set of constellation indices:

$$\mathcal{A} = \left\{ 0, 1, 2, \cdots L-1 \right\}. \tag{10.52}$$

As a result, $\beta_{bn,j}$ has the following form

$$\beta_{b_n,j} = \exp\left(i 2\pi q_{b_n,j} / L \right) \tag{10.53}$$

where $q_{b_n,j} \in \mathcal{A}$. To estimate the data symbols transmitted by the UEs $n \in \mathcal{N}$, it will suffice to estimate $q_{\tilde{b}_n,j} \; \forall \tilde{b}_n \in \tilde{\mathcal{N}}$ where $j \in \{2, 3, \ldots, |\mathcal{J}_n|\}$ since the first timeslot is reserved for the pilot symbol transmission.

Combining (10.5) and (10.53) we find

$$
\begin{aligned}
\Upsilon(n,j) &= h_{b_n}\beta_{b_n,j} \\
&= |h_{b_n}|\exp\!\left(\mathrm{i}\!\left\{\xi_{b_n}+\frac{2\pi}{L}q_{b_n,j}\right\}\right)
\end{aligned}
\tag{10.54}
$$

Here we have exploited the fact that a complex channel gain can be expressed as $h_{bn} = |h_{bn}|\exp$ $(\mathrm{i}\xi_{bn})$. We take the natural logarithm at both sides of (10.54) which yields

$$
\ln\big[\Upsilon_{n,j}\big] = \ln\big[\|h_{b_n}\|\big] + \mathrm{i}\left(\xi_{b_n}+\frac{2\pi}{L}q_{b_n,j}\right),
\tag{10.55}
$$

Separating the real and imaginary parts of the right-hand side of (10.55) gives us

$$
\mathrm{Re}\big\{\ln\big[\Upsilon(n,j)\big]\big\} = \ln\big[\|h_{b_n}\|\big],
\tag{10.56}
$$

$$
L\,\mathrm{Im}\big\{\ln\big[\Upsilon(n,j)\big]\big\} = L\xi_{b_n} + 2\pi q_{b_n,j}.
\tag{10.57}
$$

Using the fact that $q_{b_n,j} \in \mathcal{A}$, we can write

$$
\mathrm{mod}\big(L\,\mathrm{Im}\big\{\ln\big[\Upsilon(n,j)\big]\big\},\ 2\pi\big) = \mathrm{mod}\big(L\xi_{b_n},2\pi\big).
\tag{10.58}
$$

Since Υ is unknown in practice, we use instead the estimate $\tilde{\Upsilon}$ from (10.51), which contains estimation error and error caused by the perturbation of noise. The impact of these errors can be partially mitigated by averaging all the estimates for $j \in \mathcal{J}_n$ for the nth UE as shown below

$$
|\tilde{h}_{\tilde{b}_n}| = \frac{1}{|\mathcal{J}_n|}\sum_{j\in\mathcal{J}_n}\mathrm{Re}\big\{\ln\big[\tilde{\Upsilon}(n,j)\big]\big\}.
\tag{10.59}
$$

We can estimate $\mathrm{mod}\big(L\xi_{\tilde{b}_n},\ 2\pi\big)$ in a similar fashion using the definition provided in (10.58)

$$
\bar{\xi}_{\tilde{b}_n} = \frac{1}{|\mathcal{J}_n|}\sum_{j\in\mathcal{J}_n}\mathrm{mod}\big(L\,\mathrm{Im}\big\{\ln\big[\Upsilon(n,j)\big]\big\},2\pi\big).
\tag{10.60}
$$

Note that $\bar{\xi}_{\tilde{b}_n}$ is not an estimate of $\xi_{\tilde{b}_n}$, rather it is the estimation of $\mathrm{mod}\big(L\xi_{\tilde{b}_n},2\pi\big)$. Consequently, it leaves us with L possible choices of estimates of $\xi_{\tilde{b}_n}$. All of these choices can be given by the following set

$$
\mathcal{I} = \left\{\bar{\xi}_{\tilde{b}_n}+\frac{2\pi k}{L}:\ k\in\mathcal{A}\right\}.
$$

We can use the pilot symbol to resolve this ambiguity which yields

$$
\Upsilon_{n,1} = h_{b_n} = |h_{b_n}|\exp\big(\mathrm{i}\xi_{b_n}\big).
\tag{10.61}
$$

As can be seen, (10.61) can be used to get a correct estimate $\bar{\xi}_{\tilde{b}_n}$. Using nearest-neighbor policy, we can estimate $\xi_{\tilde{b}_n}$ as follows:

$$\tilde{\xi}_{\tilde{b}_n} = \arg\min_{\xi \in \mathcal{I}} \left| \exp(i\xi) - \frac{\tilde{\Upsilon}_{n,1}}{|\tilde{\Upsilon}_{n,1}|} \right|. \tag{10.62}$$

10.7.2 DATA DETECTION

As mentioned earlier, the detection of data symbols under the proposed framework is essentially equivalent to estimating $q_{b_n,j} \; \forall \; n \in \mathcal{N}$ over each $j \in \mathcal{J}_n$. Hence, data symbols for the nth UE in the jth timeslot can be estimated by solving the following nearest-neighbor problem

$$\hat{q}_{\tilde{b}_n,j} = \arg\min_{q \in \mathcal{A}} \left| \frac{\hat{\Upsilon}_{n,j}}{|\hat{\Upsilon}_{n,j}|} \exp\left(-i\hat{\xi}_{\tilde{b}_n}\right) - \exp\left(i2\pi q/L\right) \right|. \tag{10.63}$$

10.7.3 RELIABLE RECOVERY OF TRANSMITTED DATA SYMBOLS

All the UEs located near the boundary of an mMTC cell are bound to experience poor signal-to-noise ratio (SNR) because of higher path loss in an NLOS environment. On top of that, if those UEs experience deep fading or a low activation period t_a, then it becomes difficult for the detection algorithms to accurately detect their indices and data. This phenomenon is discussed in great detail in [26], where the authors propose a framework to identify the set of UEs for which the channel estimation is unreliable. However, the framewise sparse scenario considered in [26] does not directly translate to a dynamic RA scenario.

Figure 10.2 provides a visual explanation of this phenomenon, explaining the difficulties in detecting symbols from a low SNR UE. We extend the idea of coherent channel processing proposed in [26] by considering the activation period t_a instead of the number of timeslots J in an RA opportunity. This correctly quantifies the spread around the cluster center caused by noise, as can be seen in Figure 10.2. Let us first calculate the standard deviation of the estimated channel magnitude $|\tilde{h}_{\tilde{b}_n}|$ as the following:

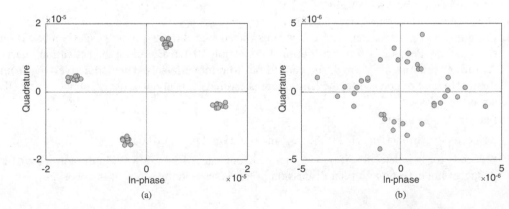

(a)   (b)

FIGURE 10.2   (a) Scatterplot of the symbols received from a reliable UE (b) Scatterplot of the symbols received from an unreliable UE [26].

$$\eta_{\tilde{b}_n} := \sqrt{\frac{1}{|\mathcal{J}_n|} \sum_{j \in \mathcal{J}_n} \left(\mathrm{Re}\left\{ \ln\left[\tilde{\Upsilon}(n,j) \right] \right\} - \| \tilde{h}_{\tilde{b}_n} \| \right)^2}. \tag{10.64}$$

Using basic geometric properties, it can be shown that the distance between two adjacent clusters can be given as $2\left|\tilde{h}_{\tilde{b}_n}\right| \sin(\pi/L)$. We can assume that the estimation error follows a normal distribution. Hence, to provide 99.7% confidence in data detection, we need

$$2\left|\tilde{h}_{\tilde{b}_n}\right| \sin\left(\pi/L\right) > 2 \times 3\eta_{\tilde{b}_n}.$$

Hence, it is possible to devise a simple threshold-based decision rule to categorize reliable and unreliable estimated UEs as follows:

$$\left|\tilde{h}_{\tilde{b}_n}\right| / \eta_{\tilde{b}_n} / > \lambda, \tag{10.65}$$

where $\lambda = 3/\sin(\pi/L)$.

Note that this framework can be extended for heterogeneous t_a without any loss of generality. Another takeaway of this analysis is that it opens up the possibility of sending ARQ-like signals in downlink to the UEs experiencing higher path loss or deep fading.

10.7.4 Summary of Proposed AUD, CE, and DD

The proposed framework begins by constructing a snapshot matrix $\bar{\Sigma}$ from the received signal $\mathbf{Y} \in \mathbb{C}^{M \times J}$. The next steps are taken in the following order:

1. The estimation of active UE set $\hat{\mathcal{N}}$ can be carried out using SPICE as outlined in Section 10.4. In each iteration, SPICE updates the covariance estimate using the steps outlined in Section 10.4. From the estimated spectrum, the estimated active UE set $\hat{\mathcal{N}}$ is formed by (10.31) via an experimentally chosen threshold.
2. Alternatively, AUD activity can be carried out using subspace estimation algorithms to solve the underlying frequency estimation problem. In that case, first the SVD of $\bar{\mathbf{S}}$ is computed. Then $|\hat{\mathcal{N}}|$ is estimated via (10.33). After that, the estimation of active UE indices $\hat{\mathcal{N}}$ is carried out via forward-backward ESPRIT.
3. Consequently, $\hat{\Upsilon}$ is found; see (10.37).
4. For a given P_{FA}, activation matrix $\mathbf{A}$ is constructed via statistical hypothesis tests based on the Neyman–Pearson criterion as outlined in Section 10.1.1. Based on the activation matrix $\mathbf{A}$, further pruning of $\hat{\mathcal{N}}$ might be carried out to remove falsely detected active UEs from the set. In case of correction, the least-squares estimate $\tilde{\Upsilon}$ is updated using refined active UE set $\tilde{\mathcal{N}}$
5. For $n \in \hat{\mathcal{N}}$, the algorithm
- proceeds to estimate $|\hat{h}_{\tilde{b}_n}|$ in (10.59), $\hat{\xi}_{\tilde{b}_n}$ in (10.62) and $\eta_{\tilde{b}_n}$ in (10.64).
- proceeds to data detection if $\eta_{\tilde{b}_n} / |\tilde{h}_{\tilde{b}_n}| > \lambda$. If not, the UEs in downlink are notified of the unsuccessful detection. Further discussion of this is beyond the scope of this work.

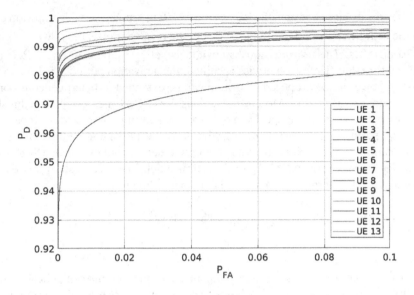

FIGURE 10.3. Individual receiver-operating characteristic (ROC) plots of the active UEs $n \in \mathcal{N}$.

10.7.5   SCOPE OF PERFORMANCE IMPROVEMENT WITH PRIOR NOISE AND CHANNEL STATISTICS

As demonstrated in Sections 10.3–10.7.3, the proposed MUD carries out active UE detection, channel estimation and data detection without any prior statistical knowledge of channels and noise, unlike most of the techniques listed in Table 10.1. In the next section, the effectiveness of ESPRIT/SPICE-based MUD will be justified by extensive numerical investigation. However, an important aspect of the proposed MUD with dynamic random access is that the symbol error rate (SER) performance of an uplink mMTC system relies on the estimate of noise variance used to carry out the null-hypothesis test demonstrated in Section 10.6. A poor estimate of noise variance leads to the inclusion of timeslots in $\mathcal{J}_n$ that consist of AWGN noise. This reduces the rate of reliable UEs demonstrated in Section 10.7.3, since it increases the standard deviation of estimated channels in Section 10.7.1. Hence, it is envisaged that prior knowledge of noise statistics will lead to enhanced SER performance. Moreover, in this work we consider homogeneous P_{FA} for all $n \in N$. It results in a different probability of detection in each UE, as shown in Figure 10.3, which plots the receiver-operating characteristic (ROC) for all $n \in \mathcal{N}$. Knowing the channel statistics of each UE $n \in N$ would also facilitate the design of UE-specific P_{FA}.

10.8   NUMERICAL ANALYSIS

10.8.1   SIMULATION SETUP

For numerical experiments, we consider an mMTC cell of 200-meter radius with a single BS at the center serving $N = 128$ UEs. We jointly model the path loss and the channel variance υ_n (in dB) using the following NLOS model (see 3GPP (release 9) for more details):

$$\upsilon_n = -128.1 - 36.7 \log_{10}\left(d_n\right), \tag{10.66}$$

where d_n denotes the distance between the BS and the nth UE in km. We consider uniform distribution of d_n within the cell, i.e., $d_n \sim \mathcal{U}[0.01, 0.2]$. To model the receiver noise, the power spectral density is set at −170 dBm/Hz. The transmission bandwidth is set at 1 MHz. A single RA opportunity consists of $J = 9$ timeslots. Unless otherwise specified, the activation period t_a is set at 5 for all $n \in \{0, 1, \cdots, N − 1\}$; however, the proposed algorithm can be extended to heterogeneous configuration of t_a without any loss of generality. Moreover, we consider Q-PSK, i.e., $L = 4$. The thresholds used for choosing peaks from the SPICE spectrum under various settings are listed in Appendix B.

The following performance metrics are taken into account in the comparative numerical analysis and bench-marking: the activity error rate (AER), the net symbol error rate (SER), and the net normalized mean squared error (NNMSE) of the estimated channels. As well as missed detections, AER also computes false alarms. NNMSE can be defined as the following:

$$\text{NNMSE} = \frac{\|\tilde{\mathbf{h}}_{\tilde{\mathcal{N}}} − \mathbf{h}_{\tilde{\mathcal{N}}}\|_2^2}{\|\mathbf{h}_{\tilde{\mathcal{N}}}\|_2^2} \tag{10.67}$$

All of these performance metrics are evaluated against a varying number of subcarriers M, activation probability p_a and transmit power γ. All of the performance curves are the ensemble average of 10,000 data points generated by Monte Carlo simulations. The performance of the ESPRIT algorithm alone is denoted ESPRIT without NPC, where NPC stands for Neyman–Pearson-criterion-aided correction. A similar convention was followed for SPICE-based AUD.

10.8.2 Simulation Results

For the study illustrated in Figure 10.4, we fix $t_a = 5$, $p_a = 0.1$, $\gamma = 20$ dBm, then vary M from 48 to 112, thus yielding an overloading from 2.67 to 1.143, It is clear from thke figure that ESPRIT with NPC outperforms its counterparts by significant margins in terms of all performance metrics under consideration. Closer observation reveals that the Neyman–Pearson-criterion-aided correction provides significant performance improvement for both ESPRIT and SPICE. Another important concern is that the performance of SPICE deteriorates significantly at low number of M, whereas ESPRIT provides relatively robust performance in that region.

In Figure 10.5 we analyze the AER, NSER and RMSE performance of different methods as the transmit power γ varies from 0 to 20 dBm for $M = 64$, $p_a = 0.1$ and $t_a = 5$. As Figure 10.5(b) shows. with increasing transmit power γ, ESPRIT provides consistent performance improvement, On the other hand, although SPICE provides significantly better performance at low transmit power, it fails to provide consistent performance improvement like ESPRIT, because SPICE's performance is also dependent on the choice of threshold (see Appendix B). Thus, the gridless version of SPICE can be regarded as a suitable candidate for MUD. However, this is subject to elaborate experiments, hence we leave this discussion for future research. Another important observation is that ESPRIT does not provide much performance improvement with Neyman–Pearson-criterion-aided correction when transmit power γ is low.

In Figure 10.6, we calculate the performance metrics for varying activation probability p_a. In this experiment we set $\gamma = 20$ dBm, $t_a = 5$ and $M = 64$. All MUD algorithms suffer from poor performance, since higher p_a translates to higher inter-cell interference among all the UEs. For high p_a, ESPRIT converges with SPICE for both NPC and non-NPC scenarios. However, in terms of SER and NNMSE, ESPRIT beats SPICE by a good margin. This is due to the fact that the statistical hypothesis testing demonstrated in Section 10.6 requires precise estimation of the noise variance σ^2. However, at higher p_a the noise variance estimated by SPICE is less accurate than that of ESPRIT [22]. This leads to increased erroneous detection of UE activity in the UAD stage, and consequently poorer SER performance for SPICE.

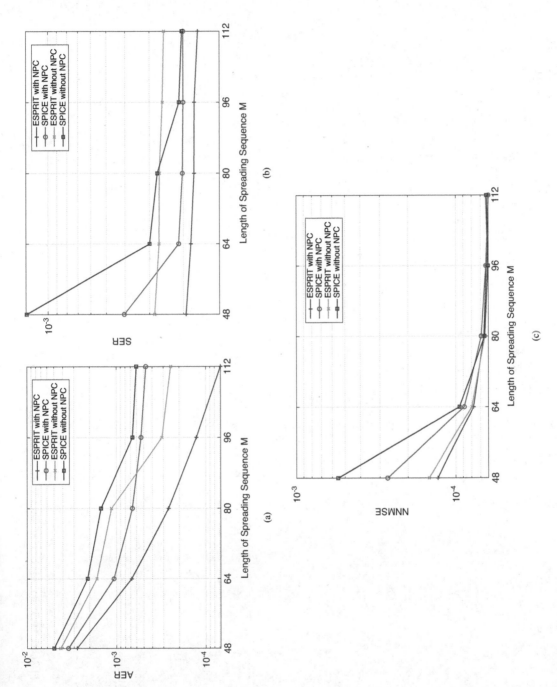

FIGURE 10.4 (a) AER (b) NSER (c) NNMSE for varying numbers of subcarriers *M*.

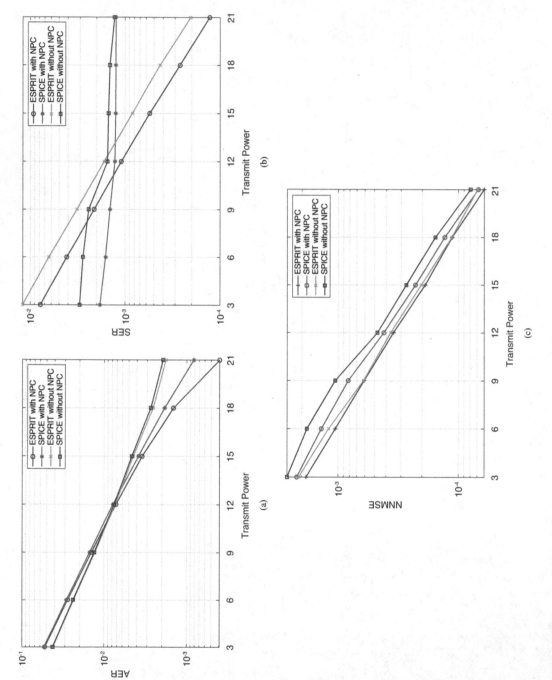

FIGURE 10.5 (a) AER (b) SER (c) NNMSE for varying transmit power γ.

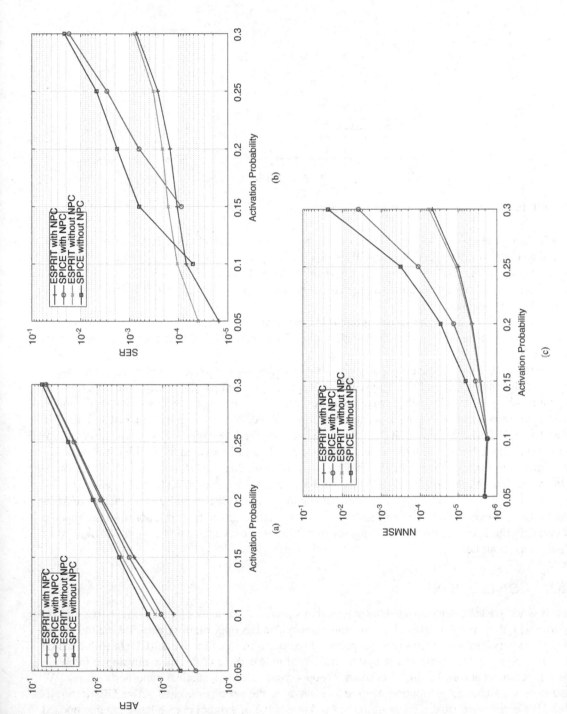

FIGURE 10.6 (a) AER (b) SER (c) NNMSE for varying activation ratio p_a.

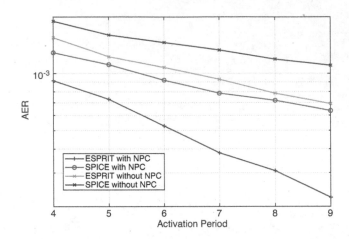

FIGURE 10.7   AER as a function of activation period t_a.

We also examine the AER performance of the proposed methods as a function of activation period t_a. For this experiment, we considered an mMTC system with $M = 64$, $p_a = 0.1$, $\gamma = 20$ dBm. Understandably, the AER performances improve substantially when t_a is high. However, it can be seen that even in a highly dynamic RA scenario, i.e., low t_a, both ESPRIT and SPICE deliver promising AER performance.

In Figure 10.8 we plot the rate of reliable UE as functions of M, γ and activation ratio p_a with $t_a = 5$. The rate of reliable UE is given as the ratio of the number of reliable UEs to the number of active UEs.

In Figure 10.9 we conduct complexity analysis by plotting the average algorithmic runtime of different algorithms for the following settings: $M = 64$, $t_a = 5$ and $\gamma = 20$ dBm. In our experiments, all the results produced by SPICE were obtained in five fixed iterations. All the numerical simulations were carried out in MATLAB on an Intel Core i7 2600, 8-core computer clocked at 3.40 GHz with 16.0 GB RAM. Figure 10.9(a) demonstrates the required average CPU time as a function of activation probability p_a. With the advantage of the acceleration provided by the FFT in the covariance matrix update equation, SPICE displays lower runtime than ESPRIT for both NPC and non-NPC scenarios. In the experiment in Figure 10.9(b), we considered an mMTC system with $M = 64$, $p_a = 0.1$, $t_a = 5$, and $\gamma = 20$ dBm. Here, we evaluate the average CPU runtime as a function of M. A similar trend is evident in this experiment, but it is noteworthy that both these algorithms are capable of carrying out MUD across a wide range of overloading in exchange for a marginal increment of algorithmic runtime.

10.9   CONCLUSIONS

In this work, we have demonstrated the potential of sinusoidal sequences for a UL grant-free mMTC system with dynamic RA, where UEs can enter and exit an RA opportunity freely. Taking the notion of user sparsity into account, we have proposed a high-performance iterative SPICE algorithm which does not require a framewise sparse system model. Simple decision rules were developed to obtain the UE activities across the timeslots of an RA opportunity, using statistical hypothesis testing. We also propose a channel estimation method to categorize the set of estimated active UEs into reliable UEs and non-reliable UEs by utilizing our knowledge of symbol constellation. Our proposed framework differs from existing works in the literature in that it does not require synchronous UE activity across the RA opportunity. While offering all these features, the proposed framework enjoys superior performance at a small computational cost. One of our major future research directions would be to consider gridless SPICE to unlock performance improvement with high transmit power.

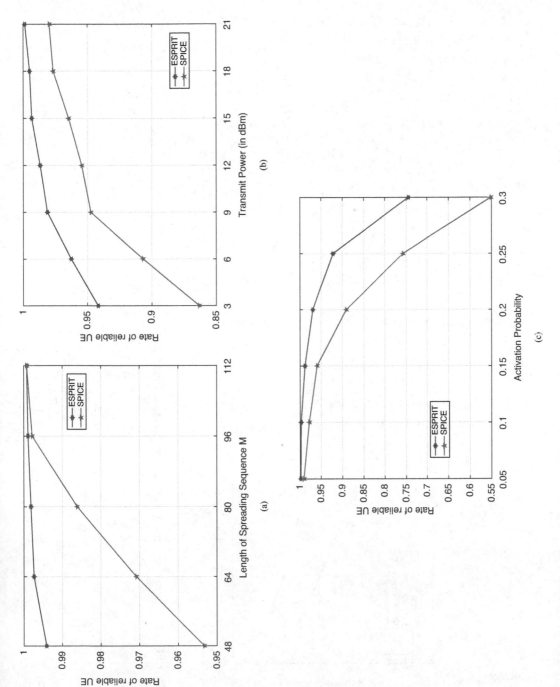

FIGURE 10.8 Rate of reliable UEs as functions of (a) M, (b) transmit power γ, and (c) p_a.

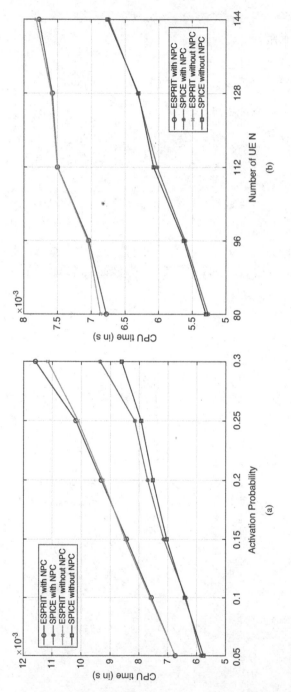

FIGURE 10.9 Average algorithmic runtime required by different methods as functions of (a) p_a, (b) N.

APPENDIX A

PROOF OF LEMMA 1

The noise-free part of $\bar{\Sigma}$ is given as

$$\bar{\Sigma} - \bar{W} = \sum_{n \in \mathcal{N}} \varkappa_n \chi_n^*,$$

This can be expressed as:

$$\bar{\Sigma} - \bar{W} = \Pi \mathbf{X} \qquad (10.68)$$

where $\Pi = \begin{bmatrix} \varkappa_{b_1} & \varkappa_{b_2} & \cdots & \varkappa_{b_{|\mathcal{N}|}} \end{bmatrix}$ and $\mathbf{X} = \begin{bmatrix} \chi_{b_1}^* & \chi_{b_2}^* & \cdots & \chi_{b_{|\mathcal{N}|}}^* \end{bmatrix}^*$ which has $|\mathcal{N}|$ rows and $\mathcal{P}$ columns. It is straightforward to validate that $\operatorname{rank}(\mathbf{X}) = |\mathcal{N}|$. Moreover, it can be seen from (10.11) that $\Pi \in \mathbb{C}^{l \times |\mathcal{N}|}$ has a Vandermonde structure. Each UE $n \in \{0, 1, \ldots, N - 1\}$ is assigned a unique sinusoidal spreading sequence which can be parameters with ω_n as described in (10.7). Hence, the Vandermonde matrix Π enjoys the following rank property: $\operatorname{rank}(\Pi) = |\mathcal{N}|$ if $l > |\mathcal{N}|$. Since $\operatorname{rank}(\Pi\mathbf{X}) \leq \min(\operatorname{rank}(\Pi), \operatorname{rank}(\mathbf{X}))$, $\operatorname{rank}(\bar{\Sigma} - \bar{W}) = |\mathcal{N}|$ for $l > |\mathcal{N}|$.

APPENDIX B

THRESHOLDS FOR SPICE

These thresholds are computed using Monte Carlo simulation. The selection criterion was a threshold that minimizes AER.

M	48	64	80	96	112
α	−26	−26	−27	−29	−30

γ	3	6	9	12	15	18	21
α	−28	−28	−28	−27	−27	−27	−26

p_a	0.05	0.1	0.15	0.2	0.25	0.3
α	−27	−26	−26	−25	−24	−24

REFERENCES

1. H. Tullberg, P. Popovski, Z. Li, M. A. Uusitalo, A. Hoglund, O. Bulakci, M. Fallgren, and J. F. Monserrat, The metis 5g system concept: Meeting the 5g requirements, *IEEE Communications Magazine*, vol. 54, no. 12, pp. 132–139, 12 2016.
2. C. Bockelmann, N. Pratas, H. Nikopour, K. Au, T. Svensson, C. Stefanovic, P. Popovski, and A. Dekorsy, Massive machine-type communications in 5G: physical and mac-layer solutions, *IEEE Communications Magazine*, vol. 54, no. 9, pp. 59–65, 9 2016.
3. K. Au, L. Zhang, H. Nikopour, E. Yi, A. Bayesteh, U. Vilaipornsawai, J. Ma, and P. Zhu, Uplink contention based SCMA for 5Gradio access, in *2014 IEEE Globecom Workshops (GC Wkshps)*, 12 2014, pp. 900–905.

4. Z. Ding, Y. Liu, J. Choi, Q. Sun, M. Elkashlan, Chih-Lin, I., and H. V. Poor, Application of non-orthogonal multiple access in LTE and 5G networks, *IEEE Communications Magazine*, vol. 55, no. 2, pp. 185–191, 2 2017.

5. Y. Liu, Z. Qin, M. Elkashlan, Z. Ding, A. Nallanathan, and L. Hanzo, Nonorthogonal multiple access for 5G and beyond, *Proceedings of the IEEE*, vol. 105, no. 12, pp. 2347–2381, 12 2017.

6. M. Shirvanimoghaddam, M. Dohler, and S. J. Johnson, Massive non-orthogonal multiple access for cellular IOT: Potentials and limitations, *IEEE Communications Magazine*, vol. 55, no. 9, pp. 55–61, 9 2017.

7. J. Hong, W. Choi, and B. D. Rao, Sparsity controlled random multiple access with compressed sensing, *IEEE Transactions on Wireless Communications*, vol. 14, no. 2, pp. 998–1010, 2 2015.

8. A. T. Abebe, and C. G. Kang, Iterative order recursive least square estimation for exploiting frame-wise sparsity in compressive sensing-based MTC, *IEEE Communications Letters*, vol. 20, no. 5, pp. 1018–1021, 2016.

9. B. Shim, and B. Song, Multiuser detection via compressive sensing, *IEEE Communications Letters*, vol. 16, no. 7, pp. 972–974, 2012.

10. B. Wang, L. Dai, T. Mir, and Z. Wang, Joint user activity and data detection based on structured compressive sensing for NOMA, *IEEE Communications Letters*, vol. 20, no. 7, pp. 1473–1476, 7 2016.

11. B. Wang, L. Dai, Y. Zhang, T. Mir, and J. Li, Dynamic compressive sensing-based multi-user detection for uplink grant-free NOMA, *IEEE Communications Letters*, vol. 20, no. 11, pp. 2320–2323, 2016.

12. C. Wei, H. Liu, Z. Zhang, J. Dang, and L. Wu, Approximate message passing-based joint user activity and data detection for NOMA, *IEEE Communications Letters*, vol. 21, no. 3, pp. 640–643, 2017.

13. Y. Du, B. Dong, Z. Chen, X. Wang, Z. Liu, P. Gao, and S. Li, Efficient multi-user detection for uplink grant-free NOMA: Prior-information aided adaptive compressive sensing perspective, *IEEE Journal on Selected Areas in Communications*, vol. 35, no. 12, pp. 2812–2828, 2017.

14. A. C. Cirik, N. Mysore Balasubramanya, and L. Lampe, Multi-user detection using admm-based compressive sensing for uplink grant-free NOMA, *IEEE Wireless Communications Letters*, vol. 7, no. 1, pp. 46–49, 2018.

15. Y. Du, B. Dong, W. Zhu, P. Gao, Z. Chen, X. Wang, and J. Fang, Joint channel estimation and multiuser detection for uplink grant-free NOMA, *IEEE Wireless Communications Letters*, vol. 7, no. 4, pp. 682–685, 2018.

16. Y. Du, C. Cheng, B. Dong, Z. Chen, X. Wang, J. Fang, and S. Li, Block-sparsity-based multiuser detection for uplink grant-free NOMA, *IEEE Transactions on Wireless Communications*, vol. 17, no. 12, pp. 7894–7909, 2018.

17. Y. Zhang, Q. Guo, Z. Wang, J. Xi, and N. Wu, Block sparse bayesian learning based joint user activity detection and channel estimation for grant-free NOMA systems, *IEEE Transactions on Vehicular Technology*, vol. 67, no. 10, pp. 9631–9640, 2018.

18. J. Zhang, Y. Pan, and J. Xu, Compressive sensing for joint user activity and data detection in grant-free NOMA, *IEEE Wireless Communications Letters*, vol. 8, no. 3, pp. 857–860, 6 2019.

19. N. Y. Yu, Multiuser activity and data detection via sparsity-blind greedy recovery for uplink grant-free NOMA, *IEEE Communications Letters*, vol. 23, no. 11, pp. 2082–2085, 11 2019.

20. J. Ahn, B. Shim, and K. B. Lee, EP-based joint active user detection and channel estimation for massive machine-type communications, *IEEE Transactions on Communications*, vol. 67, no. 7, pp. 5178–5189, 7 2019.

21. X. Miao, D. Guo, and X. Li, Grant-free NOMA with device activity learning using long short-term memory, *IEEE Wireless Communications Letters*, vol. 9, no. 7, pp. 981–984, 2020.

22. P. Stoica, P. Babu, and J. Li, SPICE: A sparse covariance-based estimation method for array processing, *IEEE Transactions on Signal Processing*, vol. 59, no. 2, pp. 629–638, 2011.

23. R. A. Horn, and C. R. Johnson, *Topics in Matrix Analysis*. Cambridge, UK: Cambridge University Press, 1991.

24. R. Roy, and T. Kailath, ESPRIT-estimation of signal parameters via rotational invariance techniques," *IEEE Transactions on Acoustics, Speech, and Signal Processing*, vol. 37, no. 7, pp. 984–995, 7 1989.

25. D. H. Johnson, Statistical signal processing, in *Encyclopedia of Biometrics*, 2009. https://cpb-us-e1.wpmucdn.com/blogs.rice.edu/dist/7/3490/files/2019/12/notes.pdf

26. S. M. Hasan, K. Mahata, and M. M. Hyder, Uplink grant-free NOMA with sinusoidal spreading sequences, *IEEE Transactions on Communications*, pp. 1–1, 2021.

27. P. Stoica, R. L. Moses et al., *Spectral analysis of signals*, Upper Saddle River, NJ: Pretnice Hall, 2005.

28. B. Ottersten, P. Stoica, and R. Roy, Covariance matching estimation techniques for array signal processing applications, *Digital Signal Processing*, vol. 8, no. 3, pp. 185–210, 1998. [Online]. Available: http://www.sciencedirect.com/science/article/pii/S1051200498903165

29. Q. Zhang, H. Abeida, M. Xue, W. Rowe, and J. Li, Fast implementation of sparse iterative covariance-based estimation for array processing, in *2011 Conference Record of the Forty Fifth Asilomar Conference on Signals, Systems and Computers (ASILOMAR)*, 2011, pp. 2031–2035.

30. M. Wax, and T. Kailath, Detection of signals by information theoretic criteria, *IEEE Transactions on Acoustics, Speech, and Signal Processing*, vol. 33, no. 2, pp. 387–392, 4 1985.

31. C. M. Hurvich, and C.-L. Tsai, Regression and time series model selection in small samples, *Biometrika*, vol. 76, no. 2, pp. 297–307, 6 1985 1989. [Online]. Available: https://doi.org/10.1093/biomet/76.2.297

32. H. Akaike, *Information Theory and an Extension of the Maximum Likelihood Principle*. New York, NY: Springer New York, 1998, pp. 199–213. [Online]. Available at: https://doi.org/10.1007/978-1-4612-1694-0_15

33. T.-J. Wu, and A. Sepulveda, The weighted average information criterion for order selection in time series and regression models, *Statistics and Probability Letters*, vol. 39, no. 1, pp. 1–10, 1998. [Online]. Available: http://www.sciencedirect.com/science/article/pii/S0167715298000030

34. P. Chen, T.. Wu, and J. Yang, A comparative study of model selection criteria for the number of signals, *IET Radar, Sonar Navigation*, vol. 2, no. 3, pp. 180–188, June 2008.

35. H. Ltkepohl, Comparison of criteria for estimating the order of a vector autoregressive process, *Journal of Time Series Analysis*, vol. 6, no. 1, pp. 35–52, 1985. [Online]. Available: https://onlinelibrary.wiley.com/doi/abs/10.1111/j.1467-9892.1985.tb00396.x

36. J. Rissanen, Modeling by shortest data description, *Automatica*, vol. 14, no. 5, pp. 465–471, 1978. [Online]. Available: http://www.sciencedirect.com/science/article/pii/0005109878900055

37. R. Bachl, The forward-backward averaging technique applied to TLS-ESPRIT processing, *IEEE Transactions on Signal Processing*, vol. 43, no. 11, pp. 2691–2699, 11 1985 1995.

38. G. Swirszcz, N. Abe, and A. C. Lozano, Grouped orthogonal matching pursuit for variable selection and prediction, in *Advances in Neural Information Processing Systems 22*, Y. Bengio, D. Schuurmans, J. D. Lafferty, C. K. I. Williams, and A. Culotta (Eds) Curran Associates, Inc., 2009, pp. 1150–1158. [Online]. Available: http://papers.nips.cc/paper/3878-grouped-orthogonal-matching-pursuit-for-variable-selection-and-prediction.pdf

11 5G-Enabled IoT
Applications and Case Studies

Mangal Singh

Symbiosis Institute of Technology, Symbiosis International (Deemed University), Pune, India

Shruti Goel

Institute of Technology, Nirma University), Ahmedabad, India

Ram Kishan Dewangan

Thapar Institute of Engineering and Technology (TIET), Patiala, India

CONTENTS

DOI: 10.1201/9781003045809-13

11.1   INTRODUCTION

The Internet of Things is one of the emerging technologies that has made people's lives easier and more comfortable. IoT is a network of heterogeneous low-powered devices connected together in such a way as to reduce human intervention. All the devices are connected to a single platform, the internet, through a communication technology. IoT has attracted a great deal of research attention due to the vast range of applications it can be used for, including smart city, smart home, wearables, smart healthcare system, agriculture and industrial automation. According to [1], IoT applications can be divided into massive and critical IoT applications. For massive IoT applications, such as are used in smart homes, smart cities and agriculture, a large number of devices need to be connected together and the end-to-end cost must be smaller. On the other hand, critical IoT applications require more reliability, ultra-low latency and higher safety levels as they are required for applications like health monitoring, industrial automation, intelligent transport systems, and so on. Among IoT design requirements are low device cost, long battery life, support for a massive number of devices, wide coverage area and security.

There are many communication technologies currently in use proposed in different research papers. These technologies can be divided into three types: short-range network, long-range network and cellular communication. Short range consists of Wi-Fi, Bluetooth, etc. Short-range networks cannot be used for connecting a large number of devices as they have a smaller coverage area. Long-range networks such as LoRa, DASH7 and SigFox can be used for wide coverage; they are among the most promising technologies for IoT and play a vital role in IoT networks. Low-power wide-area networks are widely used and highly efficient. The cellular network consists of 2G, 3G, 4G LTE and 5G. The 2G cellular network supports voice and data with about 9.6KB/s and 3G supports data up to 384KB/s. Low-power wide-area networks and 4G LTE communication technology are presently used in IoT but for machine-to-machine communications these technologies have limitations. 5G is therefore envisaged for future IoT applications.

It is anticipated that 5G will be able to overcome the challenges of scalability, privacy and security, high data rates and network management, and will allow the connection of an enormous number of devices across the globe. In this chapter different applications of IoT based on 5G have been reviewed. The chapter is structured as follows. Section 2 is divided into five subsections in which various use cases in different domains are examined in detail. Section 3 discusses various open issues and challenges for 5G-enabled IoT. The chapter concludes with Section 4.

11.2   EMERGING IoT APPLICATIONS

Advancements in technology have led to life-enhancing change. There are many applications of IoT in different fields. Each and every use case has different requirements, for example, for massive IoT applications the user equipment cost must be less. For critical IoT applications, lower latency, higher reliability and security are the most important requirements as the tiniest delay can lead to an accident. The sections below discuss some massive and critical IoT applications, with their different approaches and architectures.

11.2.1   SMART HEALTHCARE SYSTEM

The IoT for healthcare can also be referred to as 'the internet of medical things'. 5G plays a vital role in smart healthcare systems due to its high speed, large coverage area, lower latency and large bandwidth. The speed of 5G is 40 times more than that of the 4G cellular network. High data rates are required for transmitting medical images with ultra-low latency. 5G and the IoT has revolutionized the diagnosis and treatment of patients. Let us consider the case of a person from a village who has a chronic disease and has to travel to a city for treatment from a medical specialist. Going to a city every month for treatment costs a huge amount of money. It is very difficult for a poor person to afford quality treatment. But with the help of 5G and IoT, a check-up can be performed at home. IoT has made treatment possible anytime, anywhere. Among the many applications of 5G-enabled IoT are smart clothing, diagnostic services in rural areas, hospital management, robots, and the monitoring of healthcare data [2]. Many architectures have been proposed in different research papers to increase security and reliability of patient data. Wireless body area networks using sensors to detect important parameters such as sugar level or oxygen level are widely used. Other architectures like the IoT-based 5G-CCN network have been proposed to help patients in emergency and critical situations.

11.2.1.1 Sensor Node Architecture

There are three stages when the IoT can help to treat less complicated disease: sensors, trans receiver and storage for accessibility of information. When patients require a sensor that will measure oxygen or sugar levels, ECG signal or EEG signal, these signals will be transmitted to the cloud for storage and doctors can access the information and monitor whether or not their patients are taking their regular medication. If problems arise, doctors are alerted. In this way 5G-enabled IoT can help monitor health in real time, improve medical facilities in rural areas and reduce the cost of treatment.

In [2] a sensor node architecture is proposed (Figure 11.1). Wireless body sensors are needed to detect different patient parameters. The architecture consists of a transceiver, processing unit and storage unit. There are many sensor nodes in a wireless sensor network and this information is then transmitted to gateway devices which further communicate the data to doctor for treatment purposes (Figure 11.2).

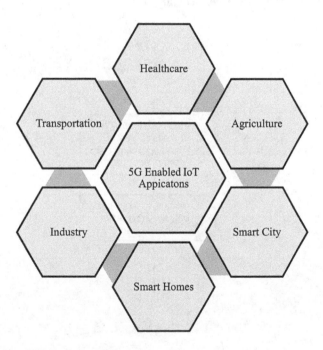

FIGURE 11.1 Use cases of 5G-enabled IoT.

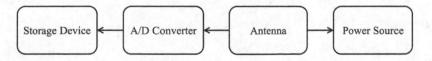

FIGURE 11.2 Sensor node architecture.

As we have seen, to make smart healthcare systems we require sensors and biomedical devices to be integrated with each other. The sensor should be comfortable and easy to handle and small in size. The antenna must be small in size so as to reduce the size of the sensor. 5G works on Gigahertz frequency so wavelength will be less and the antenna will be shorter. The information is then transmitted to the cloud via 5G connection and is stored in the cloud.

11.2.1.2 IoT-Based 5G-CCN Architecture

In [3], an IoT-based 5G-CCN network is proposed to overcome the limitations of wireless body area networks. In wireless body area networks, sensors are used to detect different parameters of a patient's body. This data is transmitted to the hospital server using communication technology. The problem with wireless body area networks is that they use an IP address for communication. The IP address has some limitations, including security issues, and sometimes communication gets lost between the hosts. To address this issue, a content-centric network (CCN) has been proposed that uses a digital signature to secure the patient's data. The CCN also has a caching facility, whereby whenever any communication takes place between the patient's server and doctor's server, the data is saved in the router. It has many advantages including scalability, security and reliability, and is useful in any emergency situation.

Consider the example of a person suffering from a disease who for some reason experiences a medical emergency. The IoT device will connect to the hospital server, but if the doctor in that hospital is not able to identify the problem, the patient may die. In IoT-based 5G-CCN, the server can connect to the nearest possible hospital and help save a life.

Figure 11.3 illustrates the 5G-CCN architecture.

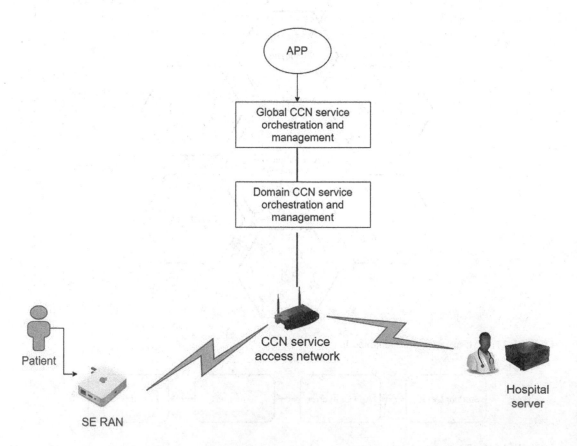

FIGURE 11.3   IoT-based 5G-CCN architecture.

In CCN architecture, the gateway is called the CCN-IoT gateway. A gateway is a device which connects two different networks. There are two CCN management services—global and domain management services. The global service orchestration and management consists of three planes: data plane, service plane and control plane. The data plane is responsible for transmitting data packets and the control plane tells the data plane which service the data has to be transmitted to. With the help of domain orchestration and management service, CCN uses cache sharing, that is, each router has the ability to store data. With the help of SE-RAN all the devices are connected to a radio access network. Some of the features of CCN like computing, caching, security, and so on, can help to support high-bandwidth 5G applications.

The protocol for this architecture consists mainly of two components: connection setup and data transmission. Initially a connection has to be set up through a request from the IoT device worn by the patient. Then the doctor's server is located, and if it receives the request, the connection setup takes place. After this data packets are transmitted between the patient's device and the doctor's server. Each and every router in the network has caching ability, which means that if another doctor wants to access the patient's medical data, then there is no need to set up a connection. The data can be accessed with the help of a data transmission component. The patient's data is stored in a table called a content store. The content store tables require information like the patient's name, medical data accessed from the patient's device, flag and sequence. Flag is used to identify whether it is a normal or emergency situation. In case of any problems, the client sends a request packet. The forwarding engine of the router checks whether the same name is present in the content store table. If the entry matches, this data packet is transmitted to the doctor's server. If the same interest request is transmitted again, it is treated as a duplicate and discarded.

In this way, an IoT-based 5G-CCN network is very beneficial and it avoids duplication of data. If the doctor is not available in one hospital, the request goes to any nearby available hospital and the patient's life may be saved. The most important advantage of this kind of architecture is that it is more secure than the wireless body area network.

11.2.1.3 Small-Cell Technology in 5G

Another very important application in healthcare systems is mobile health, one of the evolving paradigms that in future will help connect urban and rural areas. Small-cell technology is essential to fulfill the user requirements of a more complex network, high efficiency and low cost. Small cells do not require macro-cell base stations. So the cost of setting up a large number of base stations is reduced. The author in [4] has proposed an m-health use case of an ambulance which uses small-cell technology along with 5G. Let us assume there is a person Z living in a remote area and suffering from some heart-related illness. One day he has a heart attack. An ambulance arrives to take him to hospital but it is a long way from his village, there is a lot of traffic, so the ambulance is not able to reach the hospital in time and the person dies. If the ambulance had contained health-monitoring facilities like heart-beat measurement or ultrasound, perhaps these videos could be transmitted to the doctor and the patient's life saved.

Many issues arise with vehicles that are highly mobile, such as weak signal-to-noise interference ratio, or poor signal quality hindering video transmission. Small cells that can easily transmit the signal to the required server can be deployed on any vehicle (ambulance, bus). The small cell helps reduce the cost of storing large amounts of data in the cloud. Data sensed from a patient's device is transmitted to mobile and then to a small cell. The small cell consists of a server which will detect whether there is any abnormality. Any abnormal condition detected will be transmitted to the cloud, which can be accessed by a doctor. This means that only essential data needs to be stored.

Ultrasound imaging is one of the applications in the mobile ambulance system model (Figure 11.4). In [4] the authors suggest how small cells and macro cells can be combined to improve the quality of video imaging. In this model a small-cell base station is deployed in an ambulance. A transceiver is placed on the roof of the ambulance. The transceiver is used to transmit or receive data from the macro-cell link. The small-cell base station connects the ambulance crew with small-cell access

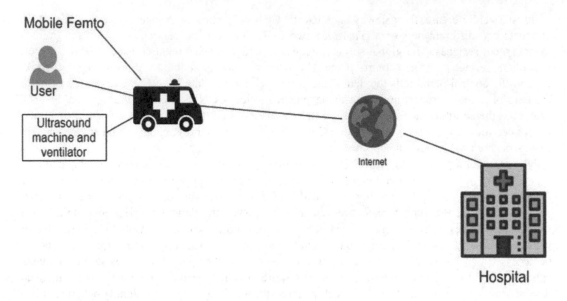

FIGURE 11.4 Ambulance scenario use case.

points. The small-cell access point and transceiver are connected through a wire. Different parameters including packet loss rate, throughput and delay were measured. The throughput in the case of the small-cell base station was more than with macro cells. This is because the signals received at the macro-cell base station are very weak, whereas strong signals are received at the micro-cell base station. There is also less delay in small-cell technology. Delay can be defined as the time it takes the data to travel from transmitter to receiver. The following analogy explains why the delay is shorter in small cell. Consider a parent queuing to get their child admitted to a school where there is only a single desk dealing with admissions and there are many parents standing in the queue. But if the admission work is divided among different persons, there will be less delay and the procedure will be completed faster. Similarly, if a large network is divided into smaller cells, then the delay will be shorter.

In this way ultrasound video streaming can be done from a mobile ambulance without compromising signal quality.

11.2.1.4 5G-Based Mobile Edge Computing

Many architectures are proposed by researchers to overcome the limitations of IoT-enabled devices, which include security and privacy, energy efficiency, resource management, low cost and battery life of IoT devices. The emergence of wireless sensor networks has eliminated the need to rush to a hospital with minor problems. The sensors sense the medical parameters of a patient, which are then transmitted to the cloud using a gateway. A gateway is a device which connects two networks. It can be a router, server or anything. In the case of smart homes/cities, the healthcare gateways are fixed. But the problem arises when the patient moves away from these fixed gateways and if any kind of emergency arises, then the data cannot be transmitted to the cloud and cannot be accessed by doctors. To overcome this limitation, the author in [5] has proposed a model in which a person's smartphone is used as a gateway. All wearable or medical IoT devices are connected to the wearer's smartphone via Bluetooth. In this way the data sensed by sensors first goes to the smartphone and from there it is transmitted to the cloud depending upon whether the data is normal or not. Smartphones are more energy efficient, and have less transmission delay and higher data rates due to the introduction of 5G. Sensing data and transmitting it to the cloud using a smartphone consumes a large amount of battery. The cost of storing all data in the cloud is also very high. The proposed

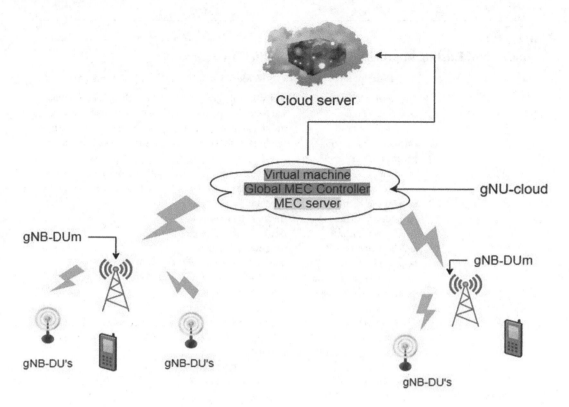

FIGURE 11.5 5G-based mobile edge computing.

approach is known as mobile edge computing or FOG computing (see Figure 11.5). The application is sent to the smartphone gateway, where it is decided whether the data is normal or requires local processing. Other information is transmitted to the MEC server where it will be executed using virtual machines present on the server.

For transmission purposes, a 5G scheme based on orthogonal frequency-division multiplexing is used with small-cell technology. The MEC server is located in the 5G base station cloud. The gNB-DUs are small-cell base stations and each small-cell BS is connected to one user device or smartphone. The gNB-DUs combined together make gNB-Dum, i.e., a macro-cell base station. The small cell is connected to the macro-cell base station via fiber cables. The MEC cloud consists of many general purpose processors (GPPs) along with virtual machines which execute applications offloaded by the smartphone gateway. The global MEC controller accepts requests for task execution and, according to the battery level, the network connection sends these tasks to virtual machines for execution. The smartphone consists of an application which identifies the tasks to be sent to the MEC cloud or for offloading purpose.

The resource monitor has the task of collecting information such as network connection state and signal speed. The purpose of the program is as follows. Information like state of charge of the battery, memory usage and CPU utilization are collected by the device profiler. A filter calculates the number of instructions executed and the execution time. All the information collected by the device profiler, resource monitor and program profiler goes to an offloading agent which further decides where to offload the task. The local execution of a task takes place if the battery SoC is less than 20% otherwise it is offloaded to the MEC cloud for execution by virtual machines. Note that with this architecture, 38% energy is saved in the mobile gateway system and the healthcare service time is reduced by 41%.

Table 11.1 provides a detailed comparison of some smart healthcare approaches.

TABLE 11.1

Comparison of Existing Approaches to Smart Health Care

Author	Year	Description	Merits	Demerits
Hina Magsi et al. [2]	2018	Proposed an architecture consisting of transceiver, processing unit and antenna for monitoring patient's health	Sensor size is small and hence low power consumption. Also comfortable to wear.	Battery charge optimization and lifetime extension of the wearable device. Security issues as data is transmitted using IP address.
Kumari Nidhi Lal et al. [3]	2017	Proposed an architecture in which content-centric network has been utilized. Patient has to send request to doctor which when accepted helps connect server of patient and doctor. It uses digital signature for data transmission.	Scalable, more secure and reliable and also helpful in any emergency situation. It has caching facility so data is saved whenever any data is transmitted from patient to doctor's server. Duplication of data is avoided.	Not suitable for denser network and mobile health care
Ikram Ur Rehman et al. [4]	2018	The author has proposed m-health use case of ambulance which uses small-cell technology along with 5G	Denser network high efficiency and less costly. Less delay and strong signal quality in case of transmitting and receiving the data.	Not cost-effective as all data need to be stored in cloud
Tshiamo Sigwele et al. [5]	2018	Mobile edge computing has been used in the proposed architecture. The smartphone is used as a gateway and from there data is transmitted to cloud.	Smartphones are more energy efficient, have less transmission delay and higher data rates due to the introduction of 5G. Less costly as only necessary data is transmitted to cloud.	Smartphone healthcare gateways can hardly cope due to limitations in terms of battery life, storage, processing power and display size. A large amount of battery will be drained which is a major concern.
Mamun, M. I., Rahman et al. [22]	2019	Proposed method in which emergency situation can be identified in a smart car using small 5G devices and passengers alerted	The system will quickly respond and alert nearby passengers, taxis and hospitals in case of any emergency. Smart cars will automatically change their position to reach the nearest hospital.	Not a cost-effective method as self-driving cars are very expensive

11.2.2 SMART AGRICULTURE

Agriculture is one of the key sectors, contributing 6.4% of the world's GDP. The Internet of Things along with 5G to make agriculture smart is envisioned as one of the most important milestones in the field of agriculture. There are many problems with traditional agriculture: farms need continuous monitoring by an individual to safeguard crops; water quantity and temperature have to be monitored in aquaculture; and large amounts of water are wasted in irrigation as the whole field does not require uniform watering. More than 69% of total water is used for agriculture purposes. Smart agriculture is seen as a solution to these challenges and can help make farming more sustainable and profitable. Smart farming will help farmers secure their crops, reduce water wastage, increase

TABLE 11.2

Comparison of Existing Approaches to Smart Agriculture

Author	Year	Description	Merits	Demerits
Ching-Kuo Hsu1 et al. [6]	2019	Image electronic fence system in which image recognition is used along with sensor fusion with 5G technology	Helpful in maintaining security of farms and avoiding crop damage by identifying unauthorized persons using image recognition with lower latency and higher accuracy	Image processing is time-consuming and is affected by climatic conditions.
Amarnathvarma Angani et al. [7]	2019	Proposed smart eel farm using IoT sensors and mobile app for real-time monitoring	Improves productivity and safety of edible fish and reduces water pollution	Challenges and issues related to the project are not discussed.
Muhammad Alam et al. [8]	2019	Project focuses on a smart irrigation system to promote rational use of water resources	Crop damage due to excessive use of water can be prevented RF energy-harvesting technology used	Issues related to real-time implementation of the project are not discussed.

productivity and profitability. Human effort may also be reduced, and efficiency of systems can be improved. There are projects or methods proposed in research papers for solving farmers' problems. In these projects different advanced technologies have been combined to make agriculture smart: for example, image recognition has been combined with 5G and IoT for secure farming. Table 11.2 compares smart agriculture approaches.

11.2.2.1 Image Electronic Fence

Farmers need help protecting their farms from unauthorized persons deliberately damaging their crops. A traditional method of ensuring the safety of the farm is by video capture from cameras placed at the entry point. The limitations of this method are that a person is needed for 24 hours to monitor people leaving/entering the farm, and that any information regarding crop damage is received by the farmer after the damage is done. In [6] an image electronic fence system is proposed in which image recognition is used along with sensor fusion with 5G technology. This technique reduces work for farmers, protects their property from unauthorized visitors, and helps reduce crop damage (Figure 11.6).

In this proposed method a BLE beacon, a wearable device, is given only to persons authorized to enter the farm. A Bluetooth receiver acts as a sensing element that identifies known individuals. The sensor identifies authorized persons by detecting the wireless signal from the BLE beacon. A surveillance camera at the entry and exit points of the farm continuously monitors the number of persons entering/leaving the farm, using image recognition technique. With the help of an object detection algorithm, bounding boxes are created for people entering/leaving the farm. The surveillance camera and sensing device are connected to the server via 5G interface so as to transmit the recorded video and sensed data with high speed and low latency. The number of people detected by video and sensed by the Bluetooth receiver are then compared. If the number of bounding boxes and number of sensed people are equal, all is well, but if not, an illegal event has occurred. These data are then transmitted to the administrator. The proposed method has an accuracy of 90%.

11.2.2.2 IoT-Based Smart Fish Agriculture

Sea farms involve a lot of manpower supplying water to the sea farm and then discharging waste water from farm to sea. The drawback with this method is that polluting waste water is directly discharged to the river or sea. The traditional method of fish cultivation is a natural water flow system, in which a control system and circulation pump are used to supply and discharge the water. The

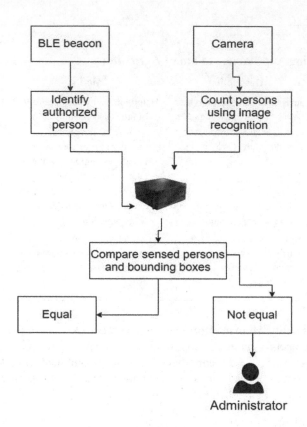

FIGURE 11.6   Image electronic fence system

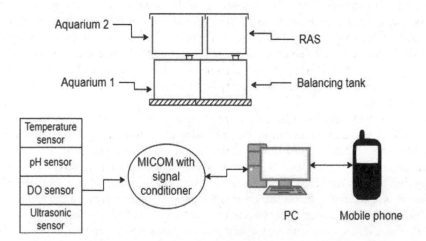

FIGURE 11.7   IoT-based smart fish farm.

authors in [7] have proposed a smart fish farm system in which ultrasonic sensors and pH sensors are used along with a mobile application. This method can be used to reduce human effort and increase productivity (Figure 11.7).

In the proposed system there are four tanks: aquarium tank 1, aquarium tank 2, balancing tank and recirculating aquaculture system (RAS) tank. They are vertically placed so as to increase fish production. The aquarium tanks 1 and 2 are used for cultivating eels. Waste water from these tanks then goes to the RAS tank, where biological and physical filtration is performed and sludge is removed. This water is then sent to the balancing tank to balance the volume of water, amount of oxygen and temperature of the water required for cultivating the eels. This balanced and purified water is then supplied to the aquarium tanks again. Three modes of operation are proposed: manual mode, remote mode and control mode. The different sensors, including ultrasonic sensor, pH sensor, DO sensor and water temperature sensors are placed in the tank along with the PID controller. Data from the different sensors goes to an Arduino microcontroller and is stored in the server. In this way different parameters like temperature, oxygen level and pH level can be controlled with a mobile application.

11.2.2.3 AREThOU5A Project

Water scarcity will be a major challenge in the coming years. More than 69% of total water is used in agriculture, and a large amount of water is wasted in irrigation. Although each area does not require same amount of water, current irrigation systems use a uniform amount of water in every part of the field, leading to inefficient use of water resources. The authors in [8] have proposed a sustainable approach for smart water irrigation using machine learning, IoT, 5G and energy harvesting. The proposed architecture (see Figure 11.8) consists of various subsystems dealing with measurement, routing, the server and the user interface. The measurement subsystem consists of different sensors for detecting soil parameters such as soil temperature, humidity, etc.

A microcontroller controls all the sensors and different power supplies including battery, solar panel and energy-harvesting module. The LoRa interface is used to transmit data to a gateway. The routing subsystem collects the sensed data and connects the measurement subsystem with the main server. The server subsystem is the place where all the data is stored permanently and all the

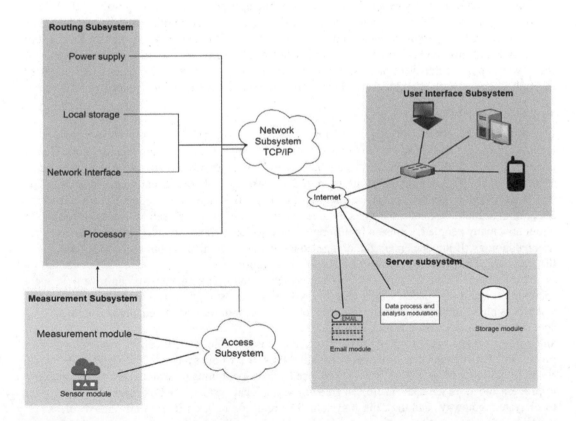

FIGURE 11.8   Architecture of AREThOU5A project.

data-processing tasks like data analysis and prediction analysis are performed. The user interface is either a mobile device such as a smartphone, or a desktop computer. In this project, the energy-harvesting module consists of an antenna to receive the RF signal and then impedance matching is performed. The RF signal is then converted to a DC signal using a rectifier. In this way IoT along with 5G can be very useful in making irrigation more sustainable.

11.2.3 Smart City

The exponential increase in population necessitates ever-growing urban areas, with problems of human safety, transportation and congestion management to be addressed. The smart city concept is envisaged as way of overcoming all these challenges. The smart city can be defined as a city where every device is connected to the internet and all services like the surveillance system, transportation system and medical system are intelligent enough to reduce human intervention. The purpose of the smart city is to make life more comfortable, improve living standards, maximize health facilities and optimize people's time. Various technologies such as block chain, information and communication technology, 5G and small-cell technology can be used to fulfill the dream of making our cities smarter. The growing population means a large amount of data will be generated every day and needs to be communicated over the internet, necessitating a vast number of connected devices, higher bandwidth and greater reliability. Some smart city use cases like medical, smart grid and surveillance systems require data to be transmitted at higher speed. This can only be achieved by employing 5G technology along with IoT. 5G has the capability to provide higher data rates and higher bandwidth and is more secure. The smart city will contribute to sustainable economic development and make life easier for people, especially the old and the disabled. A smart notification system can be developed in smart cities to notify people how far away public transport is, saving long waiting times. The smart transportation system will help reduce accidents and the smart parking system will save time spent searching for a parking space. Many accidents in the power supply system occur as a result of negligence, but this can be avoided by using a smart grid system, which will automatically cut off the power supply in the event of a fault, potentially saving lives. Research papers have suggested technologies for smart city systems, and these are reviewed in the following sections.

11.2.3.1 Smart Camera

In coastal areas beaches are mainly used for recreational and entertainment purposes by people of all ages. According to the World Health Organization, drowning is the third leading cause of accidental death and nearly 320,000 people drown annually worldwide. Our beaches therefore need to be made more secure. In [8], smart cameras are proposed to make coastal areas secure and connected. The beach is divided into three zones according to the distance from the water's edge: the dangerous red zone, orange high-alert zone and the safe green zone. An algorithm is used to process information about how many people are present in different zones, which is transmitted to lifeguards using wireless technology. If anyone enters the orange zone, the smart camera captures the image and alerts the lifeguard. The proposed architecture is shown in Figure 11.9.

With the increase in the number of private vehicles, smart parking systems can save time spent searching for parking spaces, which will also reduce fuel consumption. In a smart parking system, a sensor is placed at the center of the parking lot. The sensor detects occupied and available parking spaces and this information is transmitted to the users. In most systems there are many subsystems and a single central system. All of the data from the subsystem goes to the central system for further processing. This means that if the central system fails, the whole system will fail, in other words, reliability is reduced. In [9] a system is proposed where local units or systems have the capability to process and store the data. There is a gateway at the local system which is connected to all other local system gateways and the central system. The roadside unit, on-board units and wireless networks provide information to the local subsystem, which further processes the data. In this way the system can be made more reliable and scalable.

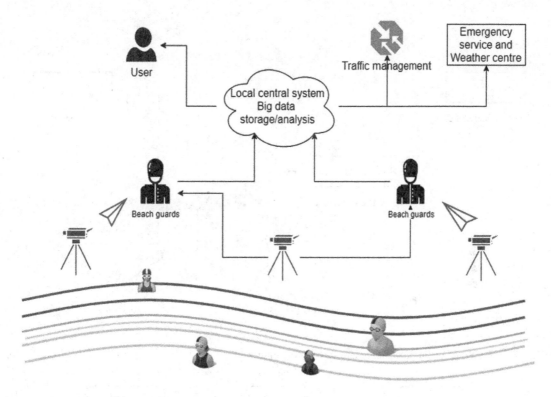

FIGURE 11.9   Beach surveillance and smart parking.

11.2.3.2   Smart Grid

In the smart grid, information and communication technology is used to manage energy production efficiently. All kinds of faults are detected and the number of interruptions can be reduced. Sometimes a huge amount of current flows through the circuit, which can lead to faults or short circuits. One of the most essential advantages of the smart grid is that it can autonomously communicate with neighboring devices and can take the required decisions. There is no need for any human intervention. When there is a power supply fault, the information is transmitted to an automated system which triggers breakers to interrupt the power supply. In this way human negligence can be avoided and accidents prevented. The distributed management approach using 5G can be employed for fault management as it requires fewer network resources and is faster, more reliable and has less latency [9]. The sensors are placed on breakers, generators and active loads. Measurement units and sensors are connected to an agent, so that each sensor and MU have one agent which can communicate with neighboring agents. In this way all agents are aware of network conditions. All the agents in a network are connected to a central subsystem through a local data aggregator. If for any reason over-current flows through a circuit, this will be detected by the measurement unit and it will trigger the agent. This agent will then communicate this measured data to the neighboring agent and the receiving agent will compare these values with the previous measurements. A fault can be avoided by triggering the breaker to the ON position. In this way there is no need to transmit the data to the central system again and again. The decision can be taken locally and hence fewer network resources are required (Figure 11.10).

11.2.3.3   Intelligent Transportation System

With increasing traffic there is a need for a smart and intelligent transportation system to reduce the number of accidents. If driverless cars are to become a reality, a technology that can assist vehicle-to-vehicle (V2V) communication is needed.

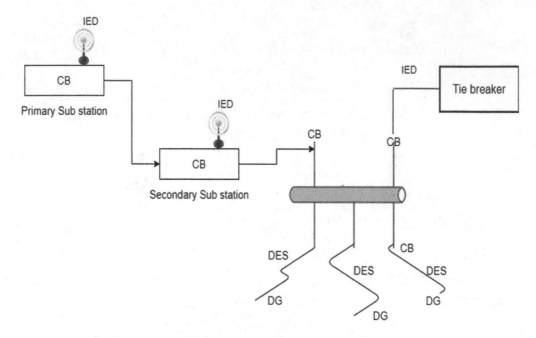

FIGURE 11.10   Distributed management system for smart grid.

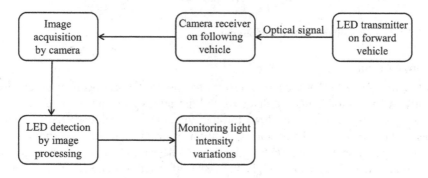

FIGURE 11.11   Vehicle-to-vehicle communication.

In [10] VLC (visible light communication) technology has been deployed, with LEDs used as brake lights or headlights on the transmitter side and a camera on the receiver side (Figure 11.11). Both sides use a processor. Vehicle information, such as speed, etc., is transmitted in the form of an optical signal. The data is collected by the processor and converted into binary form and then into one-dimensional data. This data is then transmitted to the LED. If the input is 1 then the LED is ON and if the input is 0 then the LED is OFF. A camera positioned on the receiver side and connected to a processor continuously records images and detects LEDs using image-processing techniques. The image is converted into a grayscale image and binary data can then be acquired from it. The binary data is then converted into decimal and the image can be identified.

11.2.3.4   Smart Malls

In the face of rapidly increasing urbanization, many technological and infrastructural enhancements are being introduced to make our cities smarter. The concept of smart malls covers smart parking systems, digital payment systems, smart ticket bookings, and so on. Different technologies can be employed, including IoT-enabled 5G, LoRaWAN and network function virtualization. In an

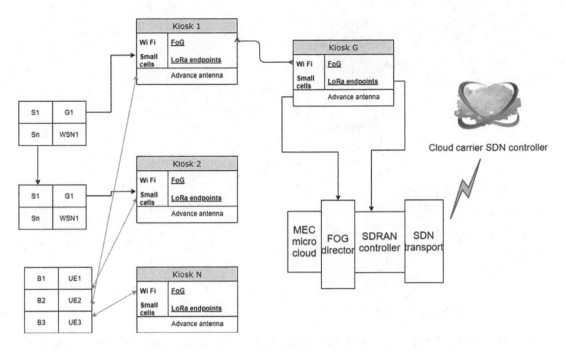

FIGURE 11.12   Architecture for smart mall.

emergency customers can be easily alerted and accidents prevented. In [11] an architecture has been proposed for converting our malls into smart and digital malls (Figure 11.12).

In this framework different wireless sensor networks, digital kiosks and gateways are deployed. A digital kiosk connected to a micro cloud through a wired medium is positioned on every floor and the high-speed connection provides visitor information displayed on a screen. Data from end-user devices are stored in the micro cloud or mobile edge computing cloud, flowing from there to the macro cloud which manages all the information. 5G can play a crucial role in the transformation into smart malls as it provides high speed and more bandwidth which may be needed in case of emergency.

11.2.3.5   Smart Surveillance System

The significant increase in the crime rate over the last 25 years poses a great challenge to our law enforcement agencies. Smart surveillance systems using the IoT can be deployed in cities to assist security officials in more rapidly and easily tracking offenders. Static or moving cameras, recorders and visual sensors can be placed at different places in a city and all these smart devices can be connected to a single platform so that they can communicate with each other. Processing and streaming large amounts of video requires high speed and very low latency. This can be achieved by employing 5G with IoT-enabled devices. Image-processing and computer-vision algorithms can also be used for automatic license plate recognition (ALPR), facial recognition and identifying brands of vehicles, all of which can assist security officials looking for offenders.

In [12] a project named 5G-CAGE proposes a solution for detecting vehicle license plate known as CODet (City Object Detection) (see Figure 11.13). A processor in each car collects all the data. The camera attached to the processor captures real-time data and the GPS coordinates of a car. This data is then transmitted to a virtual network function (VNF) through 5G mobile communication technology. Many vehicles are connected to a single VNF. This data is then processed using image-processing and computer-vision algorithms. The detected vehicle license plate, location and image is stored in a database server, which can then be searched by law enforcement officers. This project uses ALPR, OpenCV for image processing and Flask for developing web applications. The

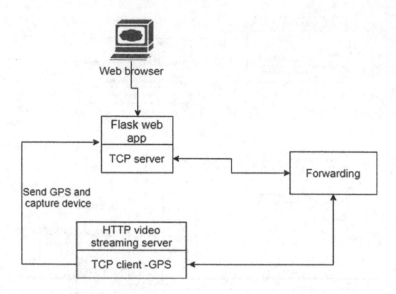

FIGURE 11.13   Smart surveillance system.

algorithm used for plate detection is a local binary pattern. Thus, IoT along with 5G and machine-learning algorithms can help secure our cities and make search operations simpler.

11.2.3.6   Smart Museums

Another use case which can be implemented in smart cities is smart museums to improve the quality of the visitor experience, helping students and others to learn about different cultural, historical and heritage artifacts in a more interesting and innovative manner. Not only students but tourists from around the world like to visit museums in order to know the history and culture of a country. The proposed scenario is shown in Figure 11.14.

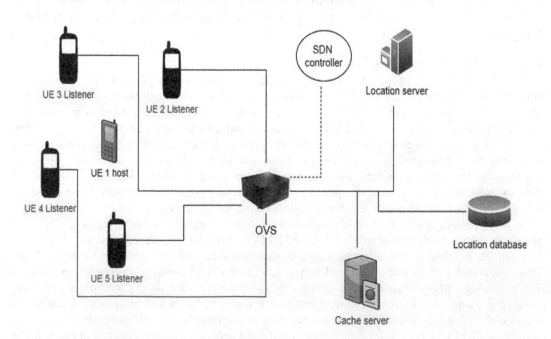

FIGURE 11.14   Museum use case.

Researchers in [13] have proposed a technology to make museum visits more memorable—a multicast sharing service (MSS) in which 5G small cells are deployed. 5G can be a key enabler as it can increase the speed of video transmission. Lecturers who want to show media content to their students create a group and become a host using MSS, then request particular media content. This request is transmitted by OVS to a cache server. Different servers are deployed for storing media content so that processing can be done locally. The requested content is searched for by the cache server and if it is not found, the request goes to another server. The host will then decide with whom the content has to be shared on the basis of location proximity to clients. Once the host declares the range, only clients or students who are in that particular area can join the group. The location is sent to the SDN controller to avoid clashes of content sharing.. In this way media content is transmitted to a particular number of clients. The proposed service depends upon the network capabilities and not the user equipment.

Existing approaches to the smart city are listed in Table 11.3

TABLE 11.3
Comparison of Existing Approaches to Smart Cities

Author	Year	Description	Merits	Demerits
Muhammad Alam et al. [8]	2017	Smart beaches along with smart parking system	Helps reduce traffic congestion and CO_2 emission and saves time searching for parking spaces. Person can be saved from drowning.	Challenges and issues regarding model data security were not discussed.
Michele Garau et al. [9]	2017	Proposed distributed management system for smart grid using 5G technology	Faults detected with more reliability and lower latency and human error avoided. Able to communicate with particular devices efficiently and selectively.	Not discussed whether the system can be used for large power network size and communication protocol should be strengthened.
Mehboob Raza Haider et al. [10]	2015	Presented an intelligent transportation system using visible light communication and image-processing techniques	LED has long life, low heat generation and high power efficiency. Many applications can be amalgamated into one camera, saving on cost and space.	Not applicable in all situations. Difficulty capturing images of fast-moving objects and obstacles can lead to signal loss. Also a very time-consuming process.
Dhanesh Raj et al. [11]	2018	Proposes an architecture for smart malls and explores key enablers: NFV-based core network, NFV-based MEC, CVC, SD-RAN etc.	Using smart mall concept visitors can be alerted in case of emergency. Help in easy navigation and localization.	Practicality of the proposed architecture was not discussed.
Pedro E. Lopez-de-Teruel et al. [12]	2019	Proposal for city object detection that enables monitoring and analysis of video streams collected from different sources of a smart city	Experimented practically and can help ensure safety of citizens. Security officers can easily detect criminals by license plate recognition.	Computational performance and accuracy of some aspects were not discussed.
Nawar Jawad et al. [13]	2019	Multicast sharing service is proposed in which 5G small cells are deployed. Teacher can cast required content to a particular group.	Adoptable in many environments and does not depend on smartphone capabilities. Provides high performance as there is zero packet loss and very high throughput.	User can only join group if the distance is less than 10 meters. Media content can be searched from selective media providers.

11.2.4 Smart Home

The smart home is seen as the most important application of evolving 5G mobile network technology. Smart meter systems, smart appliances like air conditioners, microwave ovens, washing machines, smart locks and smart lighting, and so on can work autonomously without human intervention. In some countries smart homes have been practically implemented using different communication technologies such as 4G, WLAN and LoRa, but these technologies have many limitations, including high latency, lower bandwidth, reduced security and reliability. 5G is envisioned as the most promising technology for making our homes smarter. Smart homes can not only make life easier but can also help make our environment more sustainable, contributing to resolving climate issues like increasing global warming and excessive use of energy by households. People can control their smart home appliances from anywhere at any time. The habit of not turning off lights or switching off appliances when leaving a room wastes electricity. Smart meters and smart lighting in smart homes can play a major role in reducing electricity consumption. Sensors can be positioned in rooms which can detect the presence of humans and can direct smart lighting accordingly: when someone enters the room lights will automatically be switched ON and if they leave the room, the sensor will sense that there is no one present in the room and the information to switch OFF the lights will be transmitted to the smart lighting. Smart meters help consumers to know how much energy they are consuming and can help reduce their electricity bills. Users can communicate with electricity suppliers with the help of smart meters and suppliers can notify price increases. Smoke detection is possible with the help of advances in 5G-enabled IoT.

11.2.4.1 Femtocell for Smart Home

The author in [14] has proposed a method using femtocells with WLAN devices (see Figure 11.15).

Femtocells are basically ultra-small cells which are used to enhance coverage area. Small cells can also be defined as low-power base stations that can be deployed in homes or industries to provide better quality of service. Sometimes the signals from macro base stations cannot reach indoors so small cells are employed. The signal becomes weak due to presence of obstacles between source and destination. Femtocells work as a relay to improve connectivity. In the proposed approach, the network is divided into two slices—one for the WLAN connection and the other for 5G mobile communication. The smart home router is connected to the internet service provider through an optical

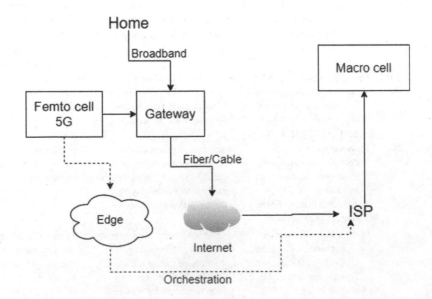

FIGURE 11.15 Femto cell architecture

cable or a wired channel. The network slicing supports both wireless LAN devices and mobile network devices. For example, if a car is in the garage, it can connect to WLAN to access information and if it is a long way from home, updates can be accessed through the 5G network. Network slicing is essential for smart homes because 5G can only support 5G devices, so there must be technology which is able to support Wi-Fi, Bluetooth and other networking technologies. Information can be accessed with minimum latency and optimal performance.

11.2.4.2 Home Energy Management System

Most energy is consumed by households. There is an alarming need to decrease energy usage to make our environment cleaner and more sustainable. The smart meter is a very appealing technology which can be employed in our homes, offering low installation cost, high reliability, secure connectivity and vast coverage area. 5G-enabled IoT can help in conserving our resources and make our world a cleaner and better place to live in.

Among the communication technologies that can be deployed, low-power wide-area network (LPWAN) uses an unlicensed spectrum for its operations which is more prone to interference. Smart meters collect the necessary information like energy consumption and notify consumers. In this way consumers can know their daily consumption and try to reduce unnecessary electricity usage. A large amount of consumption data has to be transmitted to power suppliers, so technology with higher bandwidth and low latency is required. The researchers in [15] proposed a smart and efficient energy conservation method. Meters are usually positioned in basements or storerooms, requiring a technology that can provide good coverage area and connectivity. In the proposed architecture a home gateway is employed (Figure 11.16).

11.2.4.3 Distributed Mobility Management

Some authors have used a centralized approach for connecting smart devices in homes, while others have worked on the distributed approach. Both methods have limitations. The disadvantage of the centralized approach is the risk of network failure. All the data is transmitted to one central server and if this server fails then the whole network will be affected. The distributed approach came

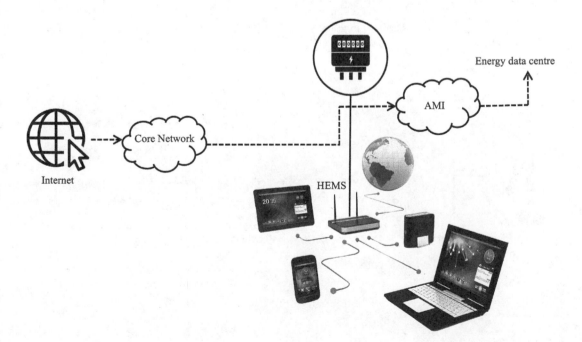

FIGURE 11.16 Smart meter

into being to overcome this limitation. In the distributed approach, different gateways are deployed which reduces the load on a single server. If one gateway fails the whole network is not at risk of failure. With distributed mobility management, the user can access smart home devices anytime and from anywhere. A large amount of data is transmitted from the user's mobile phone to the smart devices. This transmitted data is highly private and needs to be sent securely. Remotely accessed data runs a higher risk of attack, so it requires proper authentication and the number of intermediary nodes needs to be reduced. In [16] a route optimization approach has been proposed to make transmission of data more secure and less prone to attacks. In the distributed mobility management approach data is transmitted from mobile node to in-house IoT devices through two intermediary nodes, the mobile gateway and context database management. The home gateway acts as a bridge between the mobile node and IoT smart devices. By way of example, let us assume that a mobile node wants to establish a link with smart devices in their home. The link is established and data is transmitted using mobile gateway 1 and the home gateway. Now the mobile node switches to another network and a link is established between mobile gateway 2 and HGW. The data transmitted earlier to MGW 1 needs to be sent to another gateway, i.e., MGW 2, which creates the problem of indirect routing. Route optimization needs to be carried out and sensitive user information needs to be protected from external attacks.

The researchers in [16] used an authentication method for route optimization so as to make the transmission of data more secure and reliable. The authentication is required when the mobile user moves from one network to another network. The proposed approach is shown and described in Figure 11.17.To make communication more secure, an authentication method has been proposed. It consists of two phases: route optimization phase and route handover phase. This protocol makes certain assumptions: the cloud service is provided by mobile operators, the communication between

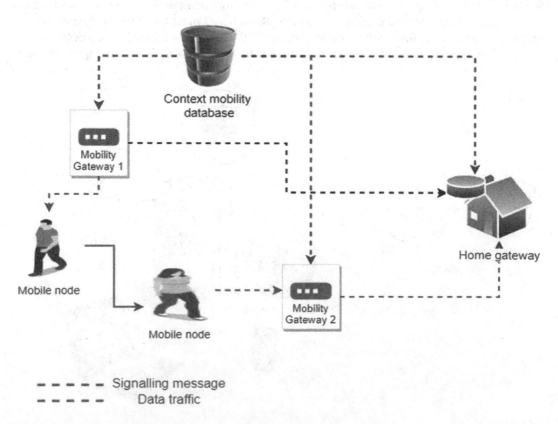

FIGURE 11.17   Distributed mobility management-based smart home system

TABLE 11.4

Comparison of Existing Approaches to Smart Homes

Author	Year	Description	Merits	Demerits
Bruno Dzogovic et al. [14]	2019	Smart home solution using femtocell home router along with network slicing	Proposed architecture can support both 5G and WLAN devices by using network slicing. System has been successfully tested at Secure 5G4IoT lab.	Unreliable Wi-Fi access can pose serious security challenge for large heterogeneous networks
Mehdi Zeinali1 et al. [15]	2017	Smart metering system using small cells in smart homes. Communication can be enabled with smart meters as well as other smart devices.	Helps consumers control resource consumption, coverage enhancement in bad coverage areas. Proposed method is also scalable.	Communication not that reliable and challenges related to model are not discussed
DAEMIN SHIN et al. [16]	2019	Security protocol for route optimization in distributed mobility management-based smart home networks	Provides perfect secrecy, mutual authentication and security from any kind of external malicious attack	Actual network performance and computation overhead is not known for real-time implementation of the model

MGWs is protected. In this method authentication and cipher keys are shared by the home gateway with the mobile gateway during initial enrollment. The home gateway initially identifies whether authentication is required between the mobile node and IoT smart devices. If authentication is required, both the keys are encrypted and transmitted to mobile gateways. The mobile gateways then decrypt the message and verify whether the received keys are valid or not. If the authentication and cipher keys are valid then communication is established between MGW and HGW. When the authentication request is received by the MGW it generates a session key.

Now if the network is changed by the mobile node from MGW1 to MGW2, then the generated session key is sent to MGW2 from CMD and the authentication process is complete. Using this protocol means that information can be exchanged more securely and effectively, preventing any external malicious attacks. The authentication and cipher keys are known only to that MGW to whom information needs to be transmitted. This protocol helps with mutual authentication, perfect forward secrecy, key exchange and defense against malicious attacks. In this way remote access can be performed by the user from anywhere at any time without worrying about data privacy and network security.

Existing approaches to the smart home are shown in Table 11.4.

11.2.5 INDUSTRIAL AUTOMATION

Industrial automation can be defined as employing different sensors, robots and actuators for sensing the information and then transmitting it to processors for performing a particular task in factories. 5G-enabled IoT can change the landscape of traditional manufacturing processes. Researchers are working on combining IoT with the 5G mobile networking technology to advance manufacturing and monitoring of products, fleet management, and so on. Industrial Revolution 4.0 will help create more business opportunities and scale up the manufacturing process. There are many factories where smart equipment has been employed or industrial automation has already taken place, but there are some fields where much of the work is done manually which leads to errors in the manufacturing process. 5G-enabled IoT can play a major role in reducing human intervention and increasing the speed of product manufacturing.

There are many factories where only product assembly is carried out. Component manufacture is outsourced to different external industries. These industries need to ensure that the components outsourced for manufacturing are of good quality so as to maintain their reputation in the market. IoT can be used to monitor component manufacturing and data collected from smart devices can be stored in the cloud to be accessed by the product company. The remote monitoring of components by the product company will help them avoid huge losses and maintain the quality of a product.

Another industry where 5G-enabled IoT can play a major role is mining. Security and safety of workers is of the utmost importance in mining. Highly toxic gases are emitted in mines which can cause lung diseases. Sensors can be placed which will gather information on toxic gas levels and inform the workers about these and any other emergency situation. 5G mobile networking offers the ultra-low latency, high reliability and high data rate which is most suitable for alerts in mining contexts. One of the main features of 5G is network function virtualization, where machines in a machine park are connected with NFVs which gather data and transmit it to the industrial cloud. Various architectures have been proposed by researchers for smart manufacturing in industries, employing different promising technologies: 5G with network function virtualization, IoT with machine learning for data analysis, distributed networking approach for avoiding server failure and reducing traffic load on a single server, and so on.

11.2.5.1   5G-Based Network Slicing

A framework is proposed in [17] for any industrial use case—product monitoring, maintenance and manufacturing (see Figure 11.18). The network is divided into different slices which will perform a particular function so as to make integration of different machines easier.

The manufacturing, networking and computing hardware, i.e., the resources required in the industrial process, come under the infrastructure plane. The virtualization plane consists of virtual network functions and routers for creating virtual industry. The slicing plane is used to represent different use cases in a process. For different functions like monitoring and maintenance, slices are generated. The function of the orchestration plane is to allocate different virtual resources to the requested industrial slice. The industrial business plane involves tools which can be used by industries to monitor the manufacturing process. The framework can be used for remote monitoring of a product. Let us assume that a product manufacturer (PM) has signed a contract with a component manufacturer (CM). A slice consisting of different VNF is allocated to PM. With the help of the orchestration system the PM can access the data regarding its own product but will not be able to access the data of the CM's other customers. The framework can also be used for equipment maintenance. Once the equipment is sold, data can be gathered by remote maintenance and any kind of optimization can be identified. The contractor sends a request to create a network slice, the request is accepted by the industry server and a network slice is created. In this way, data transmission takes place between product industry and manufacturer contractor.

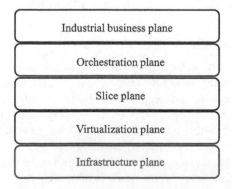

FIGURE 11.18   5G-based network slicing framework.

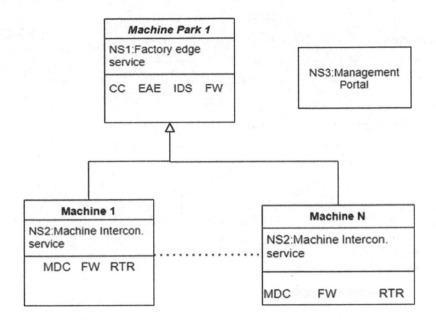

FIGURE 11.19   Smart manufacturing architecture.

11.2.5.2   Smart Manufacturing

Smart manufacturing is one of the most important use cases of industrial automation. It helps reduce human intervention and error, and also helps in scaling productivity. The authors in [18] have proposed an architecture for smart manufacturing (see Figure 11.19). It can help connect any new machine with the existing manufacturing machines in a machine park. The connection of all the machines to a single platform can help with monitoring the data of all the machines from a single location. Interconnection of the machines requires manual work which consumes a lot of time.

This proposed architecture consists of three network services: NS1, NS2 and NS3.The data is collected from sensors on manufacturing machines by NS2, which is present in each machine and translates it into messaging format. In this way each NS2 collects the data of each machine and sends it to NS1 for local processing and storage. NS2 connects each machine with the machine park's network. NS1 is present for each machine park. If a person has 50 factories all over the world, there will be 50 NS1s which will be connected to a single server. The factory edge service collects information from every machine in a single machine park and connects one machine park with other machine parks. NS1 is also connected to a factory backend cloud. The data from NS1 is transmitted to the cloud server for long-term storage and processing. When a new machine is brought into the machine park it has to be configured using the configuration screen. NS3 is present for management purposes and it generates required information. This information is then entered in the machine by the technical staff. Intrusion detection systems and firewalls can be used along with VNF. IDS will detect any malicious attack and prevent other networks from external attack. Thus, smart manufacturing architecture can integrate every machine with the rest of the machines much faster than the traditional method.

11.2.5.3   Smart Mining Industry

In some countries, the mining industry contributes significantly to the national economy, with the maximum contribution being found in lower and middle-income countries like Africa, Congo and Chile [19]. But mining involves many safety issues. Hazardous gases emitted during the mining process can cause respiratory diseases, as can inhaling coal dust. Other health issues include hearing damage due to drilling noise, chemical hazards and thermal stress. Lives can be lost in rock

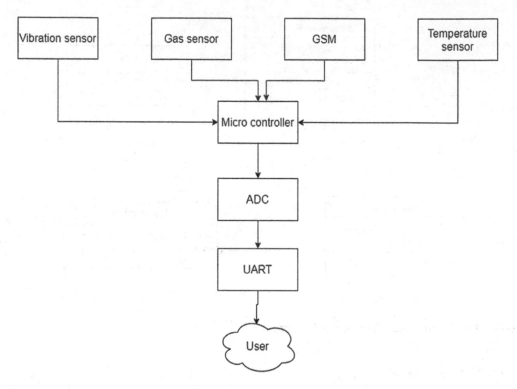

FIGURE 11.20 Smart mining technology.

falls. Miners therefore need an alert system with ultra-low latency for emergencies. IoT can play a key role in protecting them. Sensors can measure emitted toxic gases like methane and data from the sensors can then be transmitted to a cloud server where further processing can take place. 5G is highly suitable for this use case as it offers high data rate and ultra-low latency. The researchers in [20] used 5G along with machine-learning algorithms. Machine learning has high computational and data analysis abilities. The block diagram for the proposed approach is shown in Figure 11.20.

Different sensors are used to measure various toxic gas levels. An MQ2 sensor is used for measuring CO, methane and H_2, and a vibration sensor for measuring any kind of shocks. If the methane concentration is found to involve any health-related risk, the miners will be alerted. All this data is transmitted using visible light communication (VLC) and converted into a digital signal using an analog-to-digital converter. The data is sent to the cloud where machine-learning algorithms calculate whether or not the alarm needs to be raised. Only if there is a critical situation will the alarm be sounded. 5G-enabled IoT can help not only in economic development but can also help in ensuring safety of miners.

11.2.5.4 Wireless Industrial Automation

5G is seen as the future of industrial automation. Industrial automation involves sensors and actuators connected to manufacturing machines for increased efficiency and production speed. Sensor data need to be transmitted to actuators to perform a particular task. Robots can be used in the packaging industry, printing and many other industries where mass production is required. One of the use cases of 5G-enabled IoT is mentioned in [21], where a robot is used for picking and placing the pieces on a conveyor belt (Figure 11.21).

The proposed method uses sensor, robot, conveyor belt and programmable logic control. The sensor is attached to the conveyor belt which will measure the position of work pieces. This sensed data will then be transmitted to the programmable logic control (PLC) which will further process the

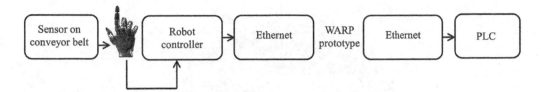

FIGURE 11.21   Wireless industrial automation.

positioning data. The PLC will identify the desired position where the work piece has to be placed and then direct the conveyor belt to reduce the speed and the robot arm to place the work piece on the conveyor belt. When this has been done, the conveyor belt will be asked to increase its speed and the robot arm will be directed to pick up the work piece. The sensors will sense the data and transmit it to the PLC using 5G mobile network technology, and the PLC will direct the robot arm to place or pick up the work pieces. Thus, wireless industrial automation can play a major role in reducing human intervention.

Table 11.5 summarizes proposed methods for smart industry.

11.3   OPEN ISSUES AND CHALLENGES

The amalgamation of IoT with 5G has received considerable research interest both from academia and industry. The combination of 5G with IoT has revolutionized the world and made people's lives easier and more comfortable. Every machine can be connected to a single platform—the internet— so that data can be easily transmitted from one device to another. The extensive literature review shows how 5G with IoT can be used to make communication more reliable and provides ultra-low latency and high data rates for data transmission. However, there are some limitations of 5G-enabled

TABLE 11.5

Comparison of Existing Approaches to Smart Industries

Author	Year	Description	Merits	Demerits
Tarik Taleb et al. [17]	2018	5G-based network slicing framework	Can be used for product monitoring, maintenance and manufacturing	Security-related issues and other challenges of the proposed model were not discussed.
Stefan Schneider et al. [18]	2019	Smart manufacturing use case to connect new machine with other machines in a machine park	Machine data can be monitored, analyzed and readjusted from a remote location. Manual error and human intervention can be reduced.	Machine data are stored in cloud servers which are prone to external malicious attacks. Cost of saving data in cloud is very high.
M S Mekala et al. [20]	2019	Light-based decision mechanism based HSM (human safety management) algorithm to evaluate and monitor abnormal conditions through an IoT sensor data	Can be used to alert miners to rise in toxic gas levels. Injuries from rock falls can be prevented.	In case of any network failure there must be an alternative which is not discussed in the paper.
Junaid Ansari et al. [21]	2017	Robot arm for picking and placing work pieces from a conveyor belt	Highly flexible, productivity is high and reliability increased	Limitations are not outlined.

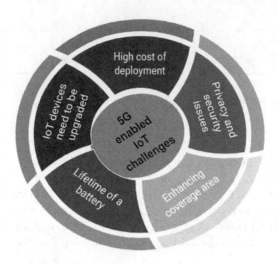

FIGURE 11.22   Research challenges in 5G-enabled IoT.

IoT which need to be overcome to achieve more secure data transfer. As every device is connected to a network, they are more vulnerable to eavesdropping, man-in-the-middle attacks, impersonation, denial of service attacks and repudiation [23]. In this section, major research challenges related to 5G-enabled IoT are identified, including the following issues:

- Some 5G architectures use a central server for data storage and processing. If a central server fails then the whole network will go down.
- With the rapid increase in battery-powered IoT devices, battery lifetime is a major challenge and energy requirements are limited.
- There are some use cases like smart health care, smart lock systems and smart industry where private data is transmitted from devices which needs to be secured.
- Integrity and data confidentiality need to be guaranteed while storing and transmitting data. Other security issues like authentication and access control, privacy, identity management and trust need to be investigated [24].
- IoT devices need to be upgraded so that they can be compatible with 5G networks [25].
- The high cost of deploying 5G network is a challenge as current IoT devices cannot be used with 5G mobile network technology. Storing data in the cloud for processing is very expensive.
- For large-scale deployment of 5G-enabled IoT, enhancing coverage area is also a great challenge.

The research challenges in 5G-enabled IoT are summarized in Figure 11.22.

11.4   CONCLUSION

This chapter has discussed different use cases in 5G-enabled IoT devices, such as smart healthcare, smart home, city, agriculture and industry. This overview sheds light on the research directions needed to make 5G network systems more reliable and secure. The various architectures proposed by researchers are summarized in tabular form. Issues and challenges related to 5G-enabled IoT are also discussed. The analysis in this chapter shows that the real spread of 5G networks and consequent service provision requires security and privacy solutions if a more robust and reliable large-scale implementation of the system is to be guaranteed. Owing to the high-end hardware requirements and network connectivity compatibility requirements, the various architectures and projects surveyed here remain a long way from reality. Future research could focus on introducing blockchain technology with 5G-enabled IoT to make data transmission more secure and prevent external malicious attacks.

5G is a very promising technology as it provides many features like low latency, high data rate, high bandwidth and more reliability. However, much research needs to be undertaken to overcome the different challenges in 5G-enabled IoT for real-time implementation of more secure and reliable systems.

REFERENCES

1. Akpakwu, G. A., Silva, B. J., Hancke, G. P., & Abu-Mahfouz, A. M. (2017). A survey on 5G networks for the Internet of Things: Communication technologies and challenges. *IEEE Access*, 6, 3619–3647.
2. Magsi, H., Sodhro, A. H., Chachar, F. A., Abro, S. A. K., Sodhro, G. H., & Pirbhulal, S. (2018). *Evolution of 5G in Internet of medical things. In 2018 International Conference on Computing, Mathematics and Engineering Technologies (Icomet)* (pp. 1–7). IEEE.
3. Lal, K. N., & Kumar, A. (2017). E-health application over 5G using Content-Centric networking (CCN). *In 2017 International Conference on IOT and Application (ICIOT)* (pp. 1–5). IEEE.
4. Rehman, I. U., Nasralla, M. M., Ali, A., & Philip, N. (2018). Small cell-based ambulance scenario for medical video streaming: a 5G-health use case. *In 2018 15th International Conference on Smart Cities: Improving Quality of Life Using ICT & IoT (HONET-ICT)* (pp. 29–32). IEEE.
5. Sigwele, T., Hu, Y. F., Ali, M., Hou, J., Susanto, M., & Fitriawan, H. (2018). Intelligent and energy efficient mobile smartphone gateway for healthcare smart devices based on 5G. *In 2018 IEEE Global Communications Conference (GLOBECOM)* (pp. 1–7). IEEE.
6. Hsu, C. K., Chiu, Y. H., Wu, K. R., Liang, J. M., Chen, J. J., & Tseng, Y. C. (2019). Design and implementation of image electronic fence with 5G technology for smart farms. *In 2019 IEEE VTS Asia Pacific Wireless Communications Symposium (APWCS)* (pp. 1–3). IEEE.
7. Angani, A., Lee, C. B., Lee, S. M., & Shin, K. J. (2019). Realization of Eel Fish Farm with Artificial Intelligence Part 3: 5G based Mobile Remote Control. *In 2019 IEEE International Conference on Architecture, Construction, Environment and Hydraulics (ICACEH)* (pp. 101–104). IEEE.
8. Alam, M., Ferreira, J., Mumtaz, S., Jan, M. A., Rebelo, R., & Fonseca, J. A. (2017). Smart cameras are making our beaches safer: A 5G-envisioned distributed architecture for safe, connected coastal areas. *IEEE Vehicular Technology Magazine*, 12(4), 50–59.
9. Garau, M., Anedda, M., Desogus, C., Ghiani, E., Murroni, M., & Celli, G. (2017). A 5G cellular technology for distributed monitoring and control in smart grid. *In 2017 IEEE international symposium on broadband multimedia systems and broadcasting (BMSB)* (pp. 1–6). IEEE.
10. Haider, M. R., & Dongre, M. M. (2015). Vehicular communication using 5G. *In 2015 International Conference on Applied and Theoretical Computing and Communication Technology (iCATccT)* (pp. 263–266). IEEE.
11. Raj, D., Lekshmi, S. S., Guruprasad, J. P., Urmila, M. S., Lakshmi, T. A., Vinod, A., & Swathi, T. V. (2018). Enabling technologies to realise smart mall concept in 5G era. *In 2018 IEEE International Conference on Computational Intelligence and Computing Research (ICCIC)* (pp. 1–6). IEEE.
12. Lopez-de-Teruel, P. E., Gil Perez, M., Garcia Clemente, F. J., Ruiz Garcia, A., & Martinez Perez, G. (2019). 5G-CAGE: A Context and Situational Awareness System for City Public Safety with Video Processing at a Virtualized Ecosystem. *In Proceedings of the IEEE/CVF International Conference on Computer Vision Workshops.*
13. Jawad, N., Salih, M., Saadoon, R., Sakat, R., Ali, K., Cosmas, J., & Zhang, Y. (2019). Indoor Unicasting/Multicasting service based on 5G Internet of Radio Light network paradigm. *In 2019 IEEE International Symposium on Broadband Multimedia Systems and Broadcasting (BMSB)* (pp. 1–6). IEEE.
14. Dzogovic, B., Santos, B., Van Do, T., Feng, B., Van Do, T., & Jacot, N. (2019). Bringing 5G Into User's Smart Home. *In 2019 IEEE Intl Conf on Dependable, Autonomic and Secure Computing* (pp. 782–787). IEEE.
15. Zeinali, M., Thompson, J., Khirallah, C., & Gupta, N. (2017). Evolution of home energy management and smart metering communications towards 5G. *In 2017 8th International Conference on the Network of the Future (NOF)* (pp. 85–90). IEEE.

16. Shin, D., Yun, K., Kim, J., Astillo, P. V., Kim, J. N., & You, I. (2019). A security protocol for route optimization in DMM-based smart home IoT networks. *IEEE Access*, 7, 142531–142550.

17. Taleb, T., Afolabi, I., & Bagaa, M. (2019). Orchestrating 5G network slices to support industrial internet and to shape next-generation smart factories. *IEEE Network*, 33(4), 146–154.

18. Schneider, S., Peuster, M., Behnke, D., Müller, M., Bök, P. B., & Karl, H. (2019). Putting 5G into Production: Realizing a Smart Manufacturing Vertical Scenario. *In 2019 European Conference on Networks and Communications (EuCNC)* (pp. 305–309). IEEE.

19. Ericsson, M., & Löf, O. (2019). Mining's contribution to national economies between 1996 and 2016. *Mineral Economics*, 32(2), 223–250.

20. Mekala, M. S., Viswanathan, P., Srinivasu, N., & Varma, G. P. S. (2019). Accurate decision-making system for mining environment using Li-Fi 5G technology over IoT framework. *In 2019 International Conference on contemporary Computing and Informatics (IC3I)* (pp. 74–79). IEEE.

21. Ansari, J., Aktas, I., Brecher, C., Pallasch, C., Hoffmann, N., Obdenbusch, M., & Gross, J. (2017). A realistic use-case for wireless industrial automation and control. *In 2017 International Conference on Networked Systems (NetSys)* (pp. 1–2). IEEE.

22. Mamun, M. I., Rahman, A., Khaleque, M. A., Mridha, M. F., & Hamid, M. A. (2019). Healthcare monitoring system inside self-driving smart car in 5g cellular network. *In 2019 IEEE 17th International Conference on Industrial Informatics (INDIN)* (Vol. 1, pp. 1515–1520). IEEE.

23. Schneider, P., & Horn, G. (2015). Towards 5G security. *In 2015 IEEE Trustcom/BigDataSE/ISPA* (Vol. 1, pp. 1165–1170). IEEE.

24. Sicari, S., Rizzardi, A., & Coen-Porisini, A. (2020). 5G in the internet of things era: an overview on security and privacy challenges. *Computer Networks*, 179, 107345.

25. Mistry, I., Tanwar, S., Tyagi, S., & Kumar, N. (2020). Blockchain for 5G-enabled IoT for industrial automation: A systematic review, solutions, and challenges. *Mechanical Systems and Signal Processing*, 135, 106382.

12 Hands-On Practice Tools for 5G and IoT

Rohan Sharma and Milind Raj

Indian Institute of Information Technology (IIIT), Bhopal,
Madhya Pradesh, India

Varun Mishra

Amity School of Engineering & Technology (ASET), Amity University,
Gwalior, India

CONTENTS

12.1 INTRODUCTION

12.1.1 5G MOBILE COMMUNICATIONS

After the evolution of 1G, 2G, 3G and 4G comes the new fifth-generation mobile network technology known as 5G. This wireless technology is capable of delivering high data speed with very low latency, enhanced efficiency and high bandwidth. It is setting a new universal standard in wireless communication by redefining a rapidly connected world [3]. It will demonstrate the most robust

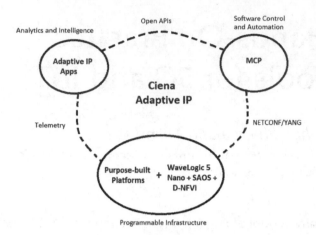

FIGURE 12.1 Adaptive IP closed-loop IP network automation, from access to metro.

technology ever experienced by the user, impacting daily life and offering a lively, seamless and more connected world [4, 6].

In today's wirelessly connected paradigm, connectivity and intelligence are predominantly associated with cloud technology. 5G, with virtually unlimited capacity, offers a new blended connectivity structure in this new virtually connected distributed computational world.

As 5G is now being deployed very rapidly, it is opening up powerful new opportunities for businesses, users and operators. The most interesting and useful applications of 5G include entertainment, connected vehicles, cloud robotics and industrial IoT. The connectivity benefits of 5G make businesses more profitable and efficient as well as giving access to more information for consumers at exceptionally fast speed.

12.1.2 INTERNET OF THINGS (IoT) TECHNOLOGY

Nowadays billions of devices all around the globe are connected to the internet, all collecting huge amounts of data and then sharing it over the network for a variety of purposes. With cheaper silicon and the ubiquity of wireless networks, it has been possible to integrate something as small as a pin to something as big as a rocket into a part of the IoT. Connecting all these devices and adding sensors to them based on specific needs adds a layer of intelligence to devices which enables them to perform intricate tasks and at the same time communicate real-time data without human involvement.

IoT has innumerable use cases, applications, standards and technologies [1]. The things and data are the starting point and essence of what IoT enables and means [2]. IoT devices share the collected sensor data by connecting to an IoT gateway from where it is either sent to the cloud/server or stored locally for analysis. These devices can communicate with other IoT-enabled devices and perform analysis on the information received from each other. The devices work with minimal or no human intervention, although we can interact with them for setting up, accessing the data or giving instructions.

IoT empowers new innovations and various improvements for consumers in business, healthcare, mobility, cities and society. Thus, the IoT has woven a fabric around us, making the world smarter and more responsive and bridging the gap between physical and digital worlds.

5G is driving a massive transformation in the world of mobile networks, bringing users a totally new connected experience, and also offers an unprecedented potential for a broad range of industries in terms of business opportunities and economic growth. As consumers, businesses, governments and many others move towards digital transformation, mobile network operators (MNOs) can precipitate this process by providing a reliable and performance-driven 5G network that is scalable, agile and cost-effective. The evolution of 5G involves constructing a network that can serve industry with new services, innovative business models and customer demands. 5G mobile networks are designed to impart ubiquitous connectivity with superfast data transfer rates and ultra-low latency.

The IoT has empowered tasks which a decade ago seemed daunting. However, despite its advantages and advancements, the IoT faces many challenges such as latency and bandwidth issues. These problems are increasing as the number of devices is increasing exponentially.

12.2 CHALLENGES AND OPPORTUNITIES

12.2.1 5G CHALLENGES AND OPPORTUNITIES

5G enables transmission of multiplex data in a blink of an eye, combining this speed with a low latency that will benefit every sector of the economy. It can be inferred that the world will eventually experience seamless AR and VR applications, remotely controlled industrial machinery and high-definition movies downloaded within seconds. It is also expected to enable future services which are unknown as of now [4].

12.2.1.1 Challenges

Spectrum availability. 5G networks require high frequencies up to 300GHz for working and delivering very fast speed i.e., up to 20 times more than the long-term evolution (LTE) network. But MNOs have to bid for the high-frequency spectrum, and cost and ability remain a struggle [8].

Less availability of 5G devices. There are still a limited number of 5G devices available in the market, other than 5G smartphones which are nowadays easily available. Some of the possible reasons include: front-end design challenges for multi-band support of frequency bands in higher and lower range; and heating issues due to increased power utilization for the transmission of high-frequency bands [8].

Low coverage area. Despite offering high speed and bandwidth, the 5G network has a limited range and requires some extra infrastructure to solve this issue. High frequency allows highly directional radio waves, which means that they are capable of being targeted or aimed. The greatest challenge for 5G antennas is that despite handling a large number of users and data, they range out within short distances and repeaters have to be installed to extend the range [9].

12.2.1.2 Opportunities

5G has three major applications in connected services: boosted and upgraded mobile broadband, massive IoT and mission-critical communications [4].

Upgraded and boosted mobile broadband. 5G mobile technology can make the smartphone experience more flawless and enhanced with new attractive experiences in the world of VR and AR, with lower latency, fast and stable data rates and less cost per bit.

Massive IoT. Massive IoT defines a category having hundreds to billions of connected devices. The main objective of these applications is to transmit and consume data rapidly and efficiently from a massive number of devices. 5G helps to achieve this task seamlessly in everything through its ability to reduce power and data costs and hence create low-cost, fluent connectivity.

Mission-critical communications. 5G has the potential to serve with new services that can transform industries by providing ultra-low latency, a credible, more available remote connection to complex machines, critical infrastructure, robots, medical machinery and future interconnected vehicles that cannot afford latency.

12.2.2 IoT CHALLENGES AND OPPORTUNITIES

Although IoT has been transforming our world in ways we never imagined by making it even more connected, even after years of development there are major issues to be solved. Over the years it has empowered many applications which were not possible before, and transformed the way we live and the way we interact with our devices. IoT is now deeply integrated into our lives: from wearables to GPS, it's everywhere.

12.2.2.1 Challenges

Some of the major challenges faced by IoT are:

Security [6]. Security is one of the most significant challenges, especially cybersecurity as most of the IoT devices are connected to digital devices over the network, making not only data but also all the connected hardware devices vulnerable to attacks [7]. Cybercriminals can directly make their way into the system and access sensitive materials through one of the many vulnerabilities present throughout the IoT. Most IoT devices have passwords set to default and old unpatched software.

Bandwidth Requirement. As the IoT market is growing at an exponential pace, connectivity has become a much bigger challenge than we perhaps anticipated. With most IoT applications becoming bandwidth intensive, for example, security camera feed, high-resolution video streaming, the current server–client model may struggle, since it utilizes a single central server for authentication and directing traffic onto multiple IoT networks. The increasing number of devices often struggle due to excessive load.

Resource Consumption. All electronic devices require energy to operate and IoT devices are a part of that. But what makes matters worse is that IoT devices have to actively monitor, transmit and receive data 24/7, and depend on other hardware like network adapters, gateways, and so on, which also need to be active all the time. This increases energy consumption significantly.

12.2.2.2 Opportunities

IoT Applications in Transportation. The transport industry is rapidly progressing towards self-driving cars. All self-driving cars will be connected to each other and will form a massive part of the IoT layer. But that's only a tiny part of the already existing IoT ecosystem in the field of transportation [10]. GPS is another IoT application that is already being used heavily by transportation companies for efficient plotting of routes. As the IoT is burgeoning in the transport industry it is highly plausible that all cars on road may eventually talk to each other, sharing relevant data, which may even automatically prevent collisions.

IoT Applications in Healthcare. We are living in the world of wearables. From a fitness band to ECG monitors, everything is part of IoT. A smartwatch can easily monitor your heart rate and derive implications from the recorded data. It saves immense time in critical situations. They also enable doctors to track their patients' health at home, decreasing resource consumption whilst providing almost real-time information.

IoT Applications in Agriculture. Agriculture has seen many advances since the advent of IoT. It makes management and monitoring climate conditions very straightforward, enabling automatic watering and fertilizer spraying. IoT sensors can sense nutrient content and soil moisture, in conjunction with weather data, and provide data for the amount of external nutrients to be added, which in turn increases production. This also prevents wastage of resources.

12.3 PARADIGMS FOR 5G AND IoT TOOLS

12.3.1 Adaptive IP

The internet has gradually evolved over several decades and has provided us with a broad outlook which has great potential to expand even further in future. This has contributed significantly to many of those protocols associated with IP no longer being required, updated, or maintained [12]. As these protocols are now obsolete, there is a move to eliminate them from present-day IP networks to reduce storage, computation and complexity costs.

12.3.1.1 Basic

Adaptive IP is part of Ciena's MCP Advanced Apps suite. It records information about network topology, latency and routing, and creates a unified virtual IP network map. This map enables a streamlined real-time view of how routing behavior affects service delivery and acts as a path computation engine (PCE) to determine which IP network parts should be optimized. So instead of struggling through a vast amount of information across multiple domains and vendors, they provide a simplified IP network [12].

This information can also be passed via open APIs to MCP to automatically configure the service and traffic flows so as to obtain an optimal performance on the IP network. Together with routing and switching platforms, these building blocks create a complete solution offering closed-loop automation that delivers optimal IP connectivity from access to metro networks [12].

This enables easy integration with existing IP networks, from access to metro, and ensures easy migration. It allows network operators to expand IP connectivity to the point where content is created and consumed by using open analytics-driven automation to significantly increase service efficiency and at the same time overcome complexity [12].

12.3.1.2 Installation Requirements

The Adaptive IP tool requires Ciena's new programmable and open routers, 5166 and 5168. These allow soft slicing and hard slicing along with segment routing and also FlexE switching for converged 4G and 5G xHaul over a common wireline infrastructure [14]. The 5166 router is used to implement Ciena's Adaptive IP with low-cost network slicing and it is also optimized for 10/25GbE to 100/200/400GbE aggregation [14]. The 5168 router supports xHaul network slicing which enables Cloud RAN architectures. It also has support for CPRI/eCPRI/RoE/ORAN, Adaptive IPTM and high-density 10/25GbE to 100/200GbE aggregation [14].

12.3.2 5G Automation

5G mobile networks are designed to impart ubiquitous connectivity with superfast data transfer rates and ultra-low latency. The networks must be credible and supple to support services provided over different slices of the network. MNOs also need to activate, modify and scale many 5G services over these slices, each modified to performance and user requirements in a very short time.

Automation plays an important role in the working of 5G networks, especially with the creation of network slices. A network slice is a logical end-to-end network defined over virtualized resources that run on top of a common physical infrastructure. Each slice is kept isolated from the others and is designed according to the specific needs of the application or user, including speed, capacity, latency, security and topology. A logically created network slice allocates resources to an application, service, users or an enterprise based on their requirements. Network slicing plays a crucial role in successful delivery of 5G services. MNOs should plan, design and activate a large number of customized network slices for their customers very quickly. They should also be able to change a slice to enable it to tackle updated requirements in a very short time [13].

12.3.2.1 Basic

The Blue Planet 5G automation solution can help MNOs in many ways, including:

- Significantly reducing the time required to bring new services to market and assure seamless performance as well
- AI- and ML-enabled automation ensure flawless deployment of new services
- Self-healing capabilities
- Efficient and optimized use of resources

- Freeing up human resources from manual and repetitive tasks associated with planning, design and deployment.

The 5G Automation Network Slicing solution provides a vendor-agnostic, intelligent automation solution that helps MNOs transition to 5G. With Blue Planet, MNOs can implement zero-touch slice lifecycle management, which consists of automating the design, creation, modification and monitoring of network slices as well as equipping underlying resources to a slice. It is developed using capabilities from the Blue Planet Intelligent Automation portfolio, a set of products already being used by leading network operators, service providers and enterprises worldwide to efficiently manage and operate multi-vendor networks. It also supports the scaling and orchestration of network resources for 5G Core, xHaul (combination of backhaul, midhaul and fronthaul), and Radio Access Network (RAN), alongside the creation and operation of network slices. It is very important for the correct positioning of VNFs inside a mobile network and allowing MNOs to increase the utilization of network resources by reallocating unutilized resources to other slices [13].

12.3.2.2 Installation Requirements
The 5G automation tool requires Ciena's new programmable and open router, 5168. These allow soft slicing and hard slicing along with segment routing and also FlexE switching for converged 4G and 5G xHaul over a common wireline infrastructure [14]. The 5168 router supports xHaul network slicing which enables Cloud RAN architectures. It also has support for CPRI/eCPRI/RoE/ORAN, Adaptive IPTM and high-density 10/25GbE to 100/200GbE aggregation [14].

12.3.3 OPENBALENA

OpenBalena is an open-source platform for deployment and management of connected devices. Devices are required to run balenaOS, a host OS created for running containers on IoT devices; balena CLI is used to control, install updates, configure application containers, check status and view logs. OpenBalena offers backend services that consist of battle-tested components, and can also securely save the information on the device, enable remote management through built-in VPN and efficiently distribute container images to devices.

12.3.3.1 Basic

- *Scalable*. Capable of managing and deploying up to a million devices.
- *Simple provisioning*. Seamless addition of new devices to pool.
- *Container based*. Advantages from the power of virtualization and optimized for the edge.
- *Built-in VPN*. Access any devices without considering their network domain.
- *Powerful API and SDK*. High, tailored scaling abilities.
- *Easy updates*. Remote update functionality for on-device software with one command.

12.3.3.2 Installation Requirements
The current release of openBalena has the following minimum version requirements:

- balenaOS v2.58.3
- balena CLI v12.38.5

If updating from previous versions of openBalena, update the balena CLI and devices to the minimum required versions for them to be completely compatible.

12.3.3.3 Commands

Consider the following steps for the deployment of an openBalena server, which along with balena CLI allows the creation and management of a group of devices running local host or on the cloud. All the devices should be in reach of openBalena.

This tutorial follows a set-up by considering two different machines:

- An openBalena server which runs Linux. The following instructions were tested with an Ubuntu 18.04 x64 server.
- A local machine which runs Windows, Linux or macOS on which the balena CLI runs (serving as a client for openBalena server). Docker should be working on the local machine to ensure that the application images can be developed and deployed to devices. The balenaEngine can also be utilized on a balenaOS device instead of Docker.

Making server ready for openBalena installation

Log in to the server through SSH and execute the commands given below and as stated in [11].

1. Installing/updating essential software:

```
$ apt-get update && apt-get install -y
build-essential git docker.io libssl-dev nodejs npm
```

2. Installing docker-compose:

```
$ curl -L
https://github.com/docker/compose/releases/download/1
.27.4/docker-compose-Linux-x86_64 -o
/usr/local/bin/docker-compose

$ chmod +x /usr/local/bin/docker-compose
```

3. Testing the docker-compose installation

```
$ docker-compose --version
```

4. Make a new user and also assign admin permissions. Also add it to the docker group:

```
$ adduser balena

$ usermod -aG sudo balena

$ usermod -aG docker balena
```

Installing openBalena on server machine

1. On server machine, log in as new user and navigate to the home directory:

```
$ su balena

$ cd ~
```

2. Then clone openBalena repository and navigate to the new directory:

```
$ git clone
https://github.com/balena-io/open-balena.git

$ cd open-balena/
```

3. Then execute the **quickstart** script as described below. This creates a new **config** directory and generates relevant configuration for the server and SSL certificates too. The user account will be automatically created using the email and password provided for communicating with the server and these credentials will be required later for logging in via the balena CLI. Change the name of the domain for the **–d** argument appropriately.

```
$ ./scripts/quickstart -U <email@address> -P
<password> -d mydomain.com
```

4. To start the openBalena server:

```
$ systemctl start docker

$ ./scripts/compose up -d
```

5. Tail the logs of the containers with:

```
$ ./scripts/compose exec <service-name> journalctl
-fn100
```

6. Change **<service-name>** for the name of any one of the services present in **compose/services.yml**; eg. **api** or **registry**
7. To stop the server:

```
$ ./scripts/compose stop
```

12.4 CONCLUSIONS

5G solves many of the issues faced by IoT, such as the requirement for superfast speeds and low latency for massive IoT. This chapter provides information on the large network infrastructure that will be required by billions of IoT devices in future. The conjunction of 5G and IoT can lead to a superfast, ultra-responsive and highly connected world which may open the door to innovation and integration of new technologies with existing systems. It will also take today's technologies even further: a robot may perform surgery remotely by following a doctor's hands; self-driving cars can communicate with each other so as to automatically prevent collisions and for efficient traffic management. Taken together, 5G and IoT can revamp present-day systems and have the potential to transform the world towards the zenith of inter-device communication. This chapter also reported different tools, such as Adaptive IP, 5G Automation and openBalena, that can provide deep analysis for IoT as well as 5G communications.

REFERENCES

1. *The Internet of Things (IoT) – essential IoT business guide*, Accessed on: Feb. 15, 2021. [Online]. Available: https://www.i-scoop.eu/internet-of-things-guide/
2. Alexander S. Gillis, *Internet of things (IoT)*, Feb. 2020. Accessed on: Feb. 15, 2021. [Online]. Available: https://internetofthingsagenda.techtarget.com/definition/Internet-of-Things-IoT

3. Qualcomm, *Unlocking the full potential of 5G*, Accessed on: Feb. 15, 2021. [Online]. Available: https://www.qualcomm.com/research/5g

4. Qualcomm, *Everything you need to know about 5G*, Accessed on: Feb. 16, 2021. [Online]. Available: https://www.qualcomm.com/5g/what-is-5g

5. Verizon, *What are 5G Ultra Wideband's benefits?* Accessed on: Feb. 16, 2021. [Online]. Available: https://www.verizon.com/about/our-company/5g/what-5g

6. Kayla Matthews, *Five major challenges when managing IoT data*, Jan. 24, 2020. Accessed on: Feb. 15, 2021. [Online]. Available: https://theiotmagazine.com/five-major-challenges-when-managing-iot-data-1bb97d890465

7. Kate Began, *5 challenges still facing the Internet of Things*, Jun. 3, 2020. Accessed on: Feb. 16, 2021. [Online]. Available: https://www.iot-now.com/2020/06/03/103228-5-challenges-still-facing-the-internet-of-things/

8. Sreenivas Midatala, *Top 7 challenges faced during 5G network deployment*, Oct. 23, 2020. Accessed on: Feb. 16, 2021. [Online]. Available: https://www.telecomlead.com/5g/top-7-challenges-faced-during-5g-network-deployment-97331

9. Ted Kritsonis, *Five of the biggest challenges facing 5G*, Jul. 2, 2020. Accessed on: Feb. 17, 2021. [Online]. Available: https://www.futurithmic.com/2020/07/02/five-biggest-challenges-facing-5g/

10. John Terra, *8 real-world IoT applications in 2020*, Sep. 14, 2020. Accessed on: Feb. 16, 2021. [Online]. Available: https://www.simplilearn.com/iot-applications-article

11. OpenBalena, *OpenBalena getting started guide*, Accessed on: Feb. 17, 2021. [Online]. Available: https://www.balena.io/open/docs/getting-started

12. Scott McFeely, *Because you asked. Adaptive IP*, Dec. 4, 2019. Accessed on: Feb. 17, 2021. [Online]. Available: https://www.ciena.com/insights/articles/because-you-asked-adaptive-ip.html

13. BluePlanet Ciena, *5G automation network slicing*, Accessed on: Feb. 17, 2021. [Online]. Available: https://www.blueplanet.com/resources/5g-automation-SB

14. Ciena, *Ciena unveils 5G innovations to fuel the next wave of mobile connectivity*, Feb. 19, 2020. Accessed on: Feb. 17, 2021. [Online]. Available: https://www.ciena.com/about/newsroom/press-releases/Ciena-Unveils-5G-Innovations-to-Fuel-the-Next-Wave-of-Mobile-Connectivity.html#:~:text=Ciena%27s%20unique%20and%20open%20approach,%2Dreliable%20Low%2DLatency%20Communications%20

Index

Page numbers in **Bold** indicate tables, page numbers in *italics* indicate figures and page numbers followed by n indicate notes.